# written in bones

# written in bones

HOW HUMAN REMAINS UNLOCK THE SECRETS OF THE DEAD

CONSULTANT EDITOR: PAUL BAHN

FIREFLY BOOKS

# A FIREFLY BOOK

Published by Firefly Books Ltd., 2003

First Printing

National Library of Canada Cataloguing in Publication Data

Written in Bones : how human remains unlock the secrets of the dead / Paul Bahn, consultant editor.

Includes bibliographical references and index.
ISBN 1-55297-685-8 (bound).—ISBN 1-55297-659-9 (pbk.)

1. Human remains (Archaeology)—2. Burial—History. 3. Tombs—History I. Bahn, Paul G.

CC77.B8W75 2003          930.1          C2002-904109-0

Publisher Cataloging-in-Publication Data  (U.S)

Bahn, Paul G.
Written in bones: how human remains unlock the secrets of the dead / Paul g. Bahn, consultant editor.— 1st ed.
[192] p. : col. ill.  ;  cm.
Includes bibliographic references and index.
Summary: How forensic archeology is used in 38 recent significant archaeological excavations.

ISBN 1-55297-685-8
ISBN 1-55297-659-9 (pbk.)

1. Archaeology—Methodology.        2. Human remains (Archaeology).
I. Title.
930.10285  21   CC73.B34  2003


Published in Canada in 2003 by
Firefly Books Ltd.
3680 Victoria Park Avenue
Toronto, Ontario M2H 3K1

Published in the United States in 2003 by
Firefly Books (U.S.) Inc.
P.O. Box 1338, Ellicott Station
Buffalo, New York 14205

This book was designed and produced by
Quintet Publishing Limited
6 Blundell Street
London N7 9 BH

Project Editor: Anna Southgate
Picture Editors: Anna Southgate and Anna Kiernan
Editor: Madeline Perri
Designer: Ian Hunt
Managing Editor: Diana Steedman
Creative Director: Richard Dewing
Publisher: Oliver Salzmann

Manufactured in Singapore by Universal Graphics Pte Ltd
Printed in China by Midas Printing

# contents

# Introduction by Paul Bahn

Human remains have always been an important source of information about many aspects of the past – and one of the most popular. When this book was being put together, no fewer than three series on British television focused on the archaeology of human remains, and the media were inundated with coverage of a necropolis of 2,000 Inca mummies found under a shantytown in Peru; the National Geographic Society screened a TV documentary on this find, which was preceded by sensational articles and photographs in numerous publications.

Mummies are big news. They sell books, such as journalist Heather Pringle's *The Mummy Congress*, an entertaining account of the often eccentric folk who devote their lives to these preserved corpses. They inspire writers of fiction – ranging from Edgar Allan Poe to Tennessee Williams, and Gustave Flaubert, who kept a mummy's foot on his desk. They arouse wonder, like the 500-year-old mummy of a Chinese of the Ming Dynasty discovered in 2001, a 60-year-old man with supple skin and mustache that was nearly a foot (30 centimeters) long; or a bejeweled elderly female Ming mummy, also found in 2001, who wore a wig to cover her baldness. They draw crowds; when the National Geographic Society put the Andean "Juanita" on show, 100,000 people turned up and waited in lines around the block. Of course they sell movies, particularly the recent *The Mummy* and *The Mummy Returns*. And it is no coincidence that the most popular computer game is called "Tomb Raider."

Mummies have even become the focus of fakery and perhaps even foul play – as in the case of the "Persian mummy" that came to light in Pakistan in October 2000 and was offered for sale on the black market for

$20 million. An elaborate fraud, the mummy, with its beaten gold decorations, was in an ornate wooden box and a stone sarcophagus. Hieroglyphics stated that the body belonged to Rhodugune or Ruduuna, the daughter of Xerxes, the great king of the Persian Empire – but Persia is not known to have practiced mummification. It all proved the work of highly skilled artisans led by a specialist scholar; executing the ruse required a goldsmith, a stonemason, a cabinetmaker, a team of embalmers and an expert on Persian history who knew its ancient language. Obviously, all this painstaking work must have cost a great deal of money. In the end, however, radiocarbon dating proved that, instead of being 2,600 years old, the mummy was at most 45 years old. It was, in fact, the mummified body of a woman who died in the mid-1990s of a broken neck and a massive back injury, and who may have been murdered or whose body may have been stolen shortly after burial.

The British Museum recently created the first virtual reality Egyptian mummy, which has allowed researchers to see detailed three-dimensional (3-D) images of an important priest called Nesperennub, who died in Thebes around 800 BC, without unwrapping him. The body was reconstructed with a medical scanner and computer graphics technology of a kind used in making the movie *The Lord of the Rings*. The 1,500 two-dimensional cross-sectional scans were pieced together by visualization software to create a complete 3-D fly-through image that can be viewed from any perspective on a computer. It resembles a hologram and shows that the priest was buried with flat, almond-shaped glass eyes, while the top of his head was covered with a strange, unfired clay bowl – for 40 years, nobody at the museum could tell from X-rays what that object could be. The pictures eventually will be used to make a model of the skull, from which an accurate reconstruction of the face will be produced. Meanwhile, visitors to the museum can use virtual reality headsets or polarized glasses to zoom in on the mummy's wrappings or even look out through its eyes. In short, in a few decades science has progressed from the destruction and unwrapping of mummies to the ability to study them thoroughly without touching them.

In another recent study, this time of a Neanderthal skull from St. Césaire, France, about 36,000 years old, scientists detected a healed fracture of the cranial vault, apparently the result of the impact of a sharp implement. This simple wound has revealed a great deal. First, in order to penetrate the bone, the blow must have been inflicted with great speed, probably by a stone tool attached to a shaft. This could not have been an accident or a wound from an animal. Second, the victim recovered, so others in the group must have provided food and shelter while the wound healed. This confirms the evidence we already have from other sites such as Shanidar (Iraq) that Neanderthals were compassionate, caring and supportive to the disadvantaged in their society. As this volume shows, this attitude continued into the late Ice Age and beyond in many cultures.

ABOVE The vast majority of human remains unearthed by the archaeologist are not mummies or preserved bodies, but skeletons; nevertheless they can provide a wealth of varied evidence, particularly in cases – as here – where they are accompanied by objects or ornaments.

Finally, in May 2002, a remarkable discovery was announced in France. Close to Clermont-Ferrand, an Iron Age tomb was unearthed containing eight people and eight small horses. Only 328 yards (300 meters) away is the impressive rampart of the unexcavated oppidum (Iron Age town) of Gondole. No tomb like this has ever been found before; multiple burials of Gauls have been discovered, and also buried horses, but never the two together, and certainly never such a spectacular array of both. The seven men and one adolescent were laid out dramatically in two rows, as were the eight horses. All 16 bodies were on their right side, head to the south, and the humans gazing east. Each man has his left arm on the man in front. What is puzzling is that there are no grave goods – no weapons, ornaments, pots, offerings or harness equipment – and no discernible cause of death. So, as yet, nobody knows if these are ritual sacrifices or victims of warfare between tribes or against Romans. What's more, two more tombs have already been detected in the immediate vicinity.

In short, there is no lack of major and exciting finds of human remains coming to light, and the extent of what we can learn about our ancestors never ceases to grow. This book intends to give the reader a mere taste of this field of study, covering the widest possible range of periods – from fossil hominids to Napoleonic troops – and of subjects, from mummies to mayhem. And yet in 50 or 100 years, even the most sophisticated of our current analyses will doubtless seem crude and primitive, such is the pace of scientific development. Our ability to read what is written in bones is constantly improving, and these new discoveries show that plenty of texts still await us.

BELOW This astounding and unique collection of skeletons, comprising eight small horses and eight men buried together, dates to the Iron Age – about 2,000 years ago – and was discovered in France in 2002.

# a way of life

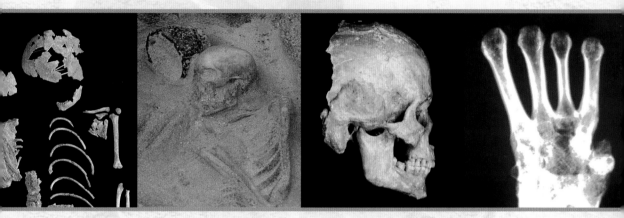

I n addition to the ongoing fascination with the peculiarities of burial ceremonies and mummification, the matter of how life was actually lived in bygone times is a compelling and informative aspect of archaeological discovery. Much can be learnt of our ancestors by examining the dietary habits and behavior of early humans.

# The Mohenjo Daro "Massacre"

A massive mudbrick platform raised the western mound at Mohenjo Daro high above the river floods that frequently threatened the city. Here on this "citadel" were located many of the city's public buildings.

In the 1920s, the discovery of ancient cities at Mohenjo Daro and Harappa in Pakistan gave the first clue to the existence more than 4,000 years ago of a civilization in the Indus Valley to rival those long known in Egypt and Mesopotamia. These cities at once struck the excavators as demonstrating an exceptional level of civic planning and amenities. The houses were furnished with brick-built bathrooms, and many had toilets. Wastewater from these was led into well-built brick sewers that ran along the center of the streets, covered with bricks or stone slabs and furnished with inspection covers. Sumps in which solid waste collected were regularly emptied. Cisterns and wells finely constructed of wedge-shaped bricks held public supplies of drinking water. Mohenjo Daro was exceptionally well furnished, probably having around 700 wells; the majority of the city's houses had a private well in addition to the public supply.

Mohenjo Daro also boasted a Great Bath on the high mound (citadel) overlooking the residential area of the city. Built of layers of carefully fitted bricks, gypsum mortar and waterproof bitumen, this basin is generally thought to have been used for ritual purification. To its north lay a series of bathrooms. At the head of the stair that gave access to the citadel was another bathroom, implying the need for purification before setting foot in the precinct. It seems this was a society in which actual and ritual purity went hand in hand and were of the greatest importance, as they are in India today.

In stark and dramatic contrast to the well-appointed houses and clean streets, the uppermost levels at Mohenjo Daro contained squalid makeshift dwellings, a careless intermingling of residential and industrial activity – and a series of sprawled skeletons lying higgledy-piggledy in the streets and houses. In a room with a public well in one area of the city were found the skeletons of two individuals who appeared desperately to have been using their last scraps of energy to crawl up the stair leading from the room to the street; the tumbled remains of two others lay nearby. Elsewhere in the area the "strangely contorted" and incomplete remains of nine individuals were found, possibly thrown into a rough pit. In a lane between two houses in another area, another six skeletons were loosely covered with earth.

Three separate groups were found in another part of the city. The first, in the melodramatically named Deadman's Lane, consisted of a thorax, upper arm and fragmentary skull of a skeleton, sprawled on its back across the narrow lane. In another lane, the remains of five more individuals lay within a thick layer of rubble, ash and debris, while in House V the remains of some 13 people were found, in contorted positions that suggested the agonies of violent death. One had a long cut across the head, perhaps from a sword blow.

These apparently shocking remains demanded an explanation. Had these people all died by violence? Some of the original excavators saw them as the victims of separate incidents of murder or armed attack. Sir Mortimer Wheeler, who excavated at Mohenjo Daro in 1950, went further and viewed them as the victims of a single massacre for which he lightheartedly claimed that the war god "Indra stands accused." He suggested that the Indus civilization, whose demise was unexplained, had fallen to an armed invasion by Indo-Aryans, nomadic newcomers from the northwest, who are thought to have settled in India during the second millennium BC. Linking the six groups of skeletons, Wheeler saw them as the defenders of the city in its final hours, bearing vivid witness to the destruction wrought by the invaders. So convincing was he, that his became the accepted version of the fate of the Indus civilization.

**BELOW** Most houses in Mohenjo Daro included a bathroom with a watertight brick-lined floor and a drain linked into the city's efficient sewerage system. Sometimes a short stair allowed someone to shower the bather from above.

But the story had many holes, and it was not long before some archaeologists, notably George Dales, began to question it. Some scholars doubt that an Indo-Aryan invasion of the subcontinent ever took place; most others agree that any influx of Indo-Aryan people, probably in small numbers, occurred after the Indus cities declined. Looking more critically at the Mohenjo Daro "massacre," there is no evidence to support the idea that these individuals were the defenders of the city. No weapons are associated with them; they lie in residential areas, not in the citadel, where the city's last stand would have occurred; there is no trace of violent destruction in the houses around them or elsewhere in the city. Nor were they cut down in flight; close examination revealed that, of all the skeletons, the bones of only two bear any trace of violence, and in both instances, the traumatic lesions had healed. In the case of the individual who had suffered a cut to the head, this had taken place at least six months before he died.

Furthermore, if these skeletons represented a massacre in the final hours of the city, they should all have belonged to the latest phase of occupation. But this was not so. Although all came from the Late Period of occupation at Mohenjo Daro, they were from different phases within it. There is a pattern, however. The skeletons all belonged to the period when civic standards had broken down. Lanes and courtyards had become

**BELOW** The "last massacre" – a dramatic element in Wheeler's theory that Indo-Aryan invaders destroyed the Indus civilization. He claimed that these sprawled remains of 13 men and women and a child lay in "attitudes resembling simultaneous death."

LEFT No cemetery has yet been found at Mohenjo Daro itself but burials at Harappa show that the Indus people usually interred their dead wearing bangles and other jewelry and furnished with some pottery.

filled with rubbish, and in many cases buildings were deliberately infilled before inferior squatter housing was erected on top. The skeletons are people from this shantytown – buried beneath house floors or dropped into abandoned streets and houses.

Recent investigations have revealed considerable evidence of flooding at Mohenjo Daro in the form of many layers of silty clay. The Indus River was prone to change its course and through the centuries moved gradually eastward, leading periodically to flooding within the bounds of the city. Indeed, the massive brick platforms on which the city is constructed and the fortifications around parts of it seem designed to provide protection against such floods. When the Saraswati River, flowing to the south of the Indus, began to dry up in the second millennium BC, one of the reasons was that its tributary, the Sutlej, was captured by the Indus system, its waters swelling the Indus and increasing its tendency to flood. The substantial infilling in the upper levels of Mohenjo Daro may have been undertaken in a desperate attempt to raise the level of houses above the advancing inundation.

Professor K. A. R. Kennedy, who has devoted himself to the study of bones from many Indian sites and periods, made an interesting discovery when he examined the "massacre" skeletons from Mohenjo Daro. They showed a high incidence of porotic hyperostosis. This condition relates to anemia due to iron deficiency or hemoglobin abnormalities. In particular, it often indicates a population suffering from thalassemia and sicklemia, commonly known as sickle-cell anemia. Although these conditions are debilitating and ultimately fatal in individuals who have inherited the condition from both parents (homozygotes), the less severe condition in heterozygotes (those who have inherited the condition from one parent only) offers some measure of protection against malaria. As a result, populations that live for a long period in a malarial region develop high concentrations of individuals with sicklemia and thalassemia. The sickle cells are less efficient in transporting oxygen around the body, leading to enlargement in the bone-forming marrow of the long bones and consequent changes in the skeleton. The effects of sicklemia and thalassemia can be distinguished from those of other anemias, affecting the frontal and parietal bones of the skull. The Mohenjo Daro skeletons

BELOW Mohenjo Daro's houses had their entrances off small lanes, away from the dust and noise of the main streets. Though now bare brick, originally the walls were plastered and probably painted with colorful designs.

exhibit thickening in these bones and large porous lesions, particularly around the eye sockets, indicating that these conditions were present.

Kennedy therefore concluded that malaria was endemic in the city, the Anopheles mosquito breeding in bodies of still water left by flooding or in the city's drainage system whenever it was not properly maintained. As public health deteriorated, a cycle of decline probably set in. Lowered standards of civic maintenance would have promoted an increase in health hazards, leading to further decline.

Porotic hyperostosis demonstrates the presence of malaria among the last inhabitants of the city. Other equally dangerous diseases would have left no trace on the bones themselves but may nevertheless have been present. The biologist Paul Ewald suggested that the efficiency of Mohenjo Daro's water supply and waste disposal system could ironically have contributed to the city's decline. Although the Indus people were careful to keep fresh water and wastewater separate, using well-engineered terra-cotta pipes and brick-lined drains, in practice their technology was inadequate to prevent seepage of water from the waste drainage system into the wells that provided the city's drinking water. When the system was not properly maintained and particularly when the city suffered from flooding, the situation became far worse. Conditions would have been ideal for the spread of water-borne diseases, especially cholera. Although cholera epidemics cannot be proved to have occurred, the circumstances make it highly probable that they did take place.

The conclusion one now reaches is that the "massacre" victims from Mohenjo Daro were indeed the victims of tragedy – but of the natural tragedy of fatal disease rather than that of human aggression. Their jumbled and often incomplete state suggests that in many cases these were not so much proper burials as attempts to dispose of bodies. Some may indeed not have been buried at all but have lain where they died until covered by falling rubble and earth from decaying buildings.

**ABOVE** Here the removal of later houses has left one of Mohenjo Daro's wells standing proud like a tower. The ubiquitous wells provided the city with abundant fresh water but also ironically promoted disease.

**BELOW** The importance of bathing and purification apparent throughout the Indus civilization is dramatically illustrated by the unique Great Bath at Mohenjo Daro. The pool's sophisticated construction ensured that it was watertight.

# You Are What You Eat

The carbon and nitrogen content of the bones of a series of child burials from the region north of Cape Town in the Western Cape Province of South Africa provides insights into the diet of the Stone Age hunter-gatherers who lived there about 2,000 years ago.

## Chemical Signatures

The chemical elements carbon and nitrogen each have a series of isotopes, atoms of the same element that have the same chemical properties but differ in mass. Carbon has three isotopes, all of which are archaeologists' friends. Many readers will be familiar with radioactive carbon 14, which occurs in minute quantities and is used for dating. However, there are also two stable forms of carbon, carbon 12 and carbon 13, which occur in a ratio of about 100:1 in the air and which can be used to reconstruct past diet.

When different kinds of plants convert carbon dioxide from the air and water into organic chemicals during the process of photosynthesis, they alter the ratio of carbon 12 to carbon 13 by discriminating against carbon 13 to different degrees. Trees, shrubs and temperate plants like wheat, which convert carbon from the air into a three-carbon molecule and hence are known as C3 plants, incorporate less carbon 13 into their tissues than do C4 plants, which make four-carbon molecules and include tropical plants like maize and sugar cane.

Marine plants, on the other hand, contain more carbon 13 than land plants. When animals and humans eat plants, or other animals, the ratio of carbon 12 to carbon 13 is altered yet again when the material is taken up by different kinds of tissues. The stable carbon isotopic content of collagen, a soft protein tissue found in human bone, can therefore indicate whether, in the course of a lifetime, the individual was in a food chain based mainly on C3 or C4 plants, or on land or marine foods, or a combination of these.

Dietary preferences can also be reflected by the ratio of nitrogen isotopes, nitrogen 14 and nitrogen 15, in bone collagen. The degree of uptake of nitrogen 15 can be used to distinguish between legume plants, like peas, that have nitrogen-fixing bacteria in their roots and those that do not, and can also be used to identify an agricultural versus a marine diet. However, interpreting the results of nitrogen ratios in bone collagen can be complicated. It was once thought that high nitrogen 15 values indicated a marine diet, but they can also reflect the consumption of foods associated with saline soils and arid environments as well as diets rich in animal foods, including those of breastfed infants.

Of course, only actual plant and animal remains can inform us exactly which species were eaten by people in the past. Plant remains in particular are rarely preserved, but the study of chemicals in bones can provide archaeologists with an additional source of information about ancient ways of life. The child burials from the Western Cape Province are a case in point.

## Coastal Dwellers

Recent excavations of the impressive Steenbokfontein Cave near the coastal town of Lambert's Bay by Antonieta Jerardino and colleagues from the University of Cape Town have revealed evidence of human occupation over the past 8,500 years. Of particular interest is a burial hollow containing the naturally desiccated partial body of an infant who died in the first few weeks after birth. Radiocarbon dating indicates the event occurred some 2,450 years ago.

### The Burial

The remains were discovered when a thick wad of grass was lifted. The infant was lying on its back in a small pit without any grave goods or any other archaeological remains with it. The face, left arm and both legs and feet were missing. Unusually, it was otherwise in an excellent state of preservation and still had soft tissue attached to parts of the skull, upper limbs, ribs and one hand. Parts of the hips are encased in tissue, while the skull is attached to the top of the spine and the arm bones are connected by tissue. The tissue is brittle and seems to comprise mainly skin, although other connective tissue such as that of the lungs may also be present. It is best preserved on the back of the body, and the excavators have suggested that bedding material on which it may have been placed could have had a fungicidal or insecticidal effect that inhibited decomposition. The dryness of the cave probably also aided preservation.

## How Old Was the Baby?

The development of the teeth suggests an age between birth and two months, while measurements of the skull and the state of other bones suggest an age of about nine fetal months, slightly behind that indicated by the teeth. The size of the bones indicates a body length of about 16½ inches (42 centimeters), which would correspond to an age of about 8.5 fetal months using European comparative material. However, given the small stature of the indigenous hunter-gatherer people of southern Africa, the infant may have been smaller than a European infant of comparable age. All considered, it seems likely that the infant did not survive long after birth.

Historical records indicate that some traditional southern African hunter-gatherer societies did not formally bury infants who had not been named. In this case, the fact that the child was carefully interred in a deliberately dug hollow and covered with grass suggests to the excavators that the child was old enough to have been named and formally

BELOW Parts of the skull of an infant burial from Steenbokfontein with soft tissue unusually well preserved by natural desiccation. The remains are stored under vacuum in a bell jar containing silica gel to keep them dry.

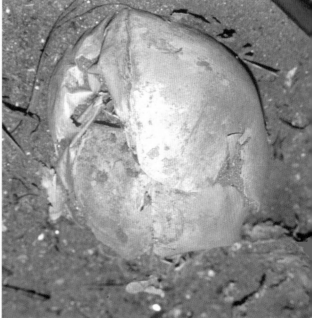

integrated into the group. It is unfortunately not possible to ascertain the sex from the bones of so young an infant, nor is the cause of death known.

### Diet

As the child was so young, most of its tissue would have formed while it was still in the mother's uterus and so essentially reflects the mother's diet. The carbon and nitrogen values obtained are typical of an individual who ate large amounts of seafood, so it seems the infant's mother heavily exploited the marine resources in the vicinity of the cave. That marine foods were important in the diets of hunter-gatherers living near the coast at the time is confirmed by the quantities of marine food remains recovered from Steenbokfontein and other archaeological sites nearby.

## Mountain People

In contrast, the skeletons of three young children recovered in 1994 and 1995 from the Pakhuis Mountains just 37¼ miles (60 kilometers) inland from Steenbokfontein indicate that people living in the interior at about the same time had diets based on land plants.

### Watervalsrivier

The burial of a young child was excavated by archaeologists from the University of Cape Town in a small rocky overhang alongside a river. Some of the bones are missing, and the site was clearly disturbed a couple of times, probably by scavenging animals. The first disturbance must have occurred very soon after the child was interred, as the lower limbs are completely absent; not even the small bones of the ankles or feet remained, which suggests that they were removed when they were still bound together by tissue. The state of development of the teeth and the bones suggests an age of just over two years at death. The body was radiocarbon dated to about 2,000 years ago.

RIGHT All that scavenging animals left of a toddler buried in a hollow cut into shallow ashy layers beneath a protective overhang in the Watervalsrivier Rock Shelter in the Pakhuis Mountains, situated north of Cape Town.

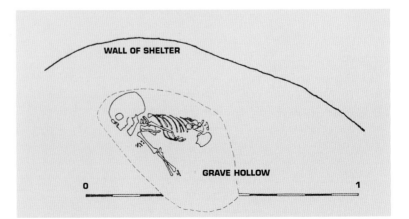

WALL OF SHELTER

GRAVE HOLLOW

0                              1

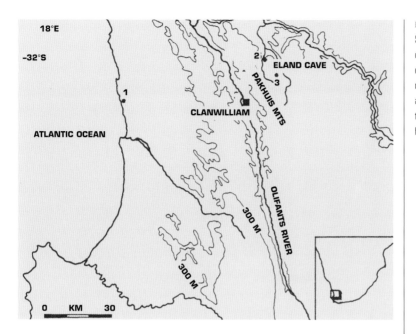

18°E

-32°S

ATLANTIC OCEAN

ELAND CAVE

PAKHUIS MTS

CLANWILLIAM

OLIFANTS RIVER

300 M

300 M

0    KM    30

LEFT The chemical content of Stone-age human bones from the coastal site of Steenbokfontein (1) contrasts with those from the inland mountain sites of Watervalsrivier (2) and Eland Cave (3), and indicates that the people in the two areas had different diets.

## Eland Cave

Two children were found in an unusual double burial in Eland Cave, a site with large numbers of rock paintings, including a beautiful frieze depicting eland antelope, near Watervalsrivier. Excavations were carried out by archaeologists from the University of Cape Town to rescue the remains that visitors had reported eroding out of shallow sediments at the back of the cave. The skeletons were reburied after study. Radiocarbon dating indicates the children died about 2,150 years ago.

Both skeletons were incomplete, but much dried soft tissue was preserved, including the complete right foot of one of the individuals. The burial hollow was lined with thick wads of grass and the bodies covered with grass and sand. Fragments of leather were found beneath the bodies, indicating that they had been placed on animal skin. Two of the fragments have stitched joining pieces; the inner side of one was decorated with a rectangular design cut into the leather, while the other had a series of parallel or cross-hatched lines scraped into it. No grave goods were found.

The teeth recovered from the burial known as Eland Cave 1 indicate an age of three years plus or minus six months. This is consistent with the state of development of the back bones, which suggests an age between two and seven years. Loose teeth associated with the second child, Eland Cave 2, indicate an age of about six to seven years, while the stage of development of its back bones suggests a minimum age of seven years.

The bones provide no clue about the cause of the children's death. Their long bones were X-rayed to look for lines indicating interruptions in growth. Large numbers of such lines are thought to reflect chronic ill

ABOVE This leather fragment shows a rectangular pattern incised on outer side of the skin. The rectangles are approximately ¼-inch (4–5 mm) across.

19

RIGHT An unusual double-child burial from Eland Cave in the Pakhuis Mountains. L1 to L5 are fragments of leather and the large stone above the legs of one of the children is an ocher-stained grindstone.

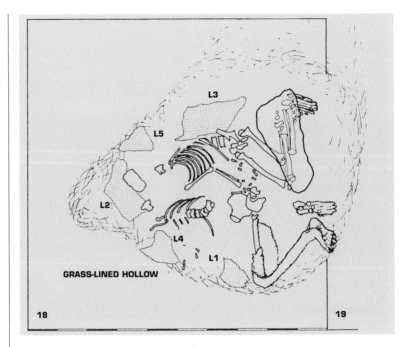

health and commonly form between the ages of two and five. Their presence on some of the Eland Cave bones suggests that they may have experienced periods of growth interruption, but not regularly.

## Diet

The carbon and nitrogen isotope readings on the children's bones indicate that their diet was based on land plants. This is similar to results from adult skeletons from other sites in the mountains and contrasts strongly with the seafood-based diet of people living at the coast at sites like Steenbokfontein at roughly the same time. The mountain people presumably visited the coast, which was not far away, but the isotopic data suggest this must have occurred only occasionally. This in turn raises questions about the mobility and ranges of the hunter-gatherers of the time as well as about the nature of the interaction between the mountain and the coastal peoples.

The child burials from the Western Cape Province of South Africa show that human bones can provide clues not only about human origins, relationships, appearance, age, sex and causes of death; they also contain chemical elements which can provide archaeologists with fascinating new windows on the diet and nutrition of past populations.

# The Lapedo Child

The Lapedo Valley is located in central Portugal, about 87 miles (140 kilometers) north of Lisbon, near the city of Leiria. In November 1998, archaeologists came here to check reports – which proved true – that prehistoric rock paintings had been found; in the course of their investigations they also discovered a limestone rock shelter, the Lagar Velho site. The upper two or three meters of its fill had been bulldozed away in 1992 by the landowner, which left a hanging remnant of sediment in a fissure along the back wall, but this contained such a density of Upper Paleolithic stone tools, bones and charcoal that it was clear that Lagar Velho had been an important occupation site.

Subsequent excavations confirmed this, producing radiocarbon dates of 23,170 to 20,220 years ago. More intriguing, however, was the discovery of human bones and red ocher at the eastern end of the shelter, at ground level; this turned out to be a child's grave, the only Paleolithic burial ever found in the Iberian Peninsula. The bulldozer had crushed the skull but, by a miracle, had missed the rest of the body by ¾ inch (2 centimeters). A rescue excavation was carried out at Christmas, initially in conditions

**ABOVE** Two views of the child's remains in the course of excavation. This apparent hybrid has become one of the most controversial discoveries in the history of archaeology. Note its proximity to the wall of the shelter.

of great secrecy to avoid any dangers to the unique find from media exposure. The work was difficult because tiny plant roots had penetrated the spongy bones. Sieving of the disturbed sediments led to the recovery of 160 cranial fragments, which constitute about 80 percent of the total skull. The postcranial skeleton was intact. It belonged to a child of three-and-a-half to five years old (on the basis of the development of the dentition) who was carefully and deliberately buried in an extended position in a shallow pit so that the head and feet were higher than the hips (this is how the skull was hit by the bulldozer). The hands were close to the hips. The body had been placed on a burned Scots pine branch, probably in a hide covered in red ocher. The ocher was particularly thick around the head and stained the upper and lower surfaces of the bones.

The body was accompanied by a complete rabbit carcass between its legs and some remains of red deer, notably pelvis bones, by its head and feet. There were also six ornaments, also stained with red ocher: four perforated canines from four different red deer (two male and two female) and two periwinkle shells from the Atlantic (*Littorina obtusata*). The deer teeth were associated with the child's skull fragments, so were probably part of a headdress. One perforated shell was complete and found in situ over the child's left shoulder, near the cervical vertebrae, so is thought to be a pendant.

The worn traces of suspension on all these ornaments prove that they were not just funerary attire but had been much used. Unfortunately, the burial had no other archaeological context – although identical perforated teeth and shells were found in Lagar Velho's occupation layers – its age had to be confirmed by radiocarbon dating. Analysis of charcoal fragments from the grave, as well as rabbit and other herbivore bones, produced results between 25,000 and 24,000 years ago. Unfortunately, the lack of collagen in the child's bones means that we can learn nothing of its bone chemistry or DNA.

## A Morphological Hybrid

But while this child's burial was of great importance in itself, another feature proved of crucial significance. When the bones were unearthed, it was noticed that the proportions of its robust lower limbs were not those of modern humans but rather resembled those of a Neanderthal. Yet the two human forms are not thought to have coexisted later than 28,000 years ago in Iberia. How could the child have features of both forms?

For example, Neanderthals had poorly developed chins and a jaw that sloped backward, whereas modern humans have a pronounced chin; the child has a chin, but that whole region of the jaw slopes back, an archaic feature. The overall shape of its skull is modern, as is the shape of its inner ear (as revealed by CAT scans) – also the teeth, which are fairly small, especially the front ones. However, even though the head, painstakingly reconstructed from the scattered, crushed and deformed fragments, looks modern, one detail was detected – a pitting in the occipital region (known as *semispinalis capitis fossae*) – that is a diagnostic and genetic trait of Neanderthals alone.

The obvious deduction – that this child is a morphological mosaic, a hybrid of Neanderthals and modern humans – has led to a bitter debate among specialists. Those who cannot accept such a concept have dismissed the skeleton as "a chunky modern child," yet this term itself highlights the dilemma, because early modern human skeletons are not chunky at all; they have long limbs. It is Neanderthals who were chunky. It has also been suggested that the child's proportions reflect a cold adaptation in a modern human population – an apparent advantage in conserving body heat; this part of the world was very cold for a few millennia before the child's time, with icebergs off the coast of Portugal. However, adult skeletons from the same period, which are known from Wales, Moravia and even northern Russia, all display the tropical body proportions of Africans, with no cold adaptations.

The unavoidable conclusion, therefore, on present evidence, is that the Lapedo child represents a hybrid of Neanderthals and anatomically modern humans. It is a modern child with genetically inherited Neanderthal traits – which means that the last Neanderthals of Iberia (and doubtless other parts of Europe) contributed to the gene pool of subsequent populations.

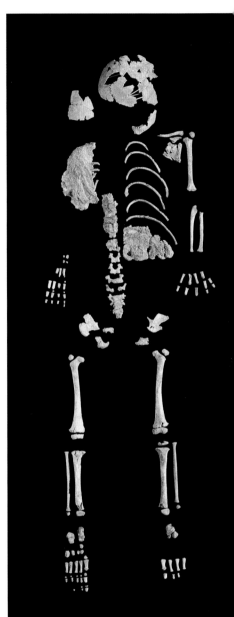

**ABOVE** The skeleton of the Lapedo child showing its "chunky" lower limbs which have already become the crux of a major debate – namely, how much did Neanderthals contribute to our gene pool?

# The Moundville Dwarf Burials

P rehistoric dwarf burials are extremely rare (see p. 148). The Moundville specimens represent the only recorded case in North America of multiple dwarf burials occurring at a single site. Although they were excavated in the 1930s, they still provide important information about the prehistoric inhabitants of Moundville.

Moundville is situated in west-central Alabama, on the south bank of a bend in the Black Warrior River. Comprising 22 earthen temple and burial mounds spread over approximately 160 acres, this is one of the largest sites of the Mississippian culture in the southeastern United States.

Moundville developed out of a small pre-Mississippian village about AD 700, with the subsequent Mississippian period extending from AD 1050 to 1550. From AD 1250 to 1450, this was the paramount power in the region, with a resident population approximating 3,000 people. After 1450, the community went into decline and was abandoned by the time of De Soto's mid-sixteenth-century exploration of the area for Spain. Excavations have taken place at Moundville repeatedly since 1866.

Although not actually interred together, the two dwarf burials were found close to each other in the eastern portion of the site. The female was found first, in 1934, and the male five years later, in 1939. The female was found face down in a tightly flexed position on top of a normal-sized skeleton. The male dwarf was also found face down, but only in a partially flexed position. Both individuals had been buried in shallow surface graves in this habitation area of residential surface structures that appears to have been occupied sometime after AD 1400. The location

RIGHT The site of Moundville. Both dwarf burials were found in the upper right portion of the open site. Most of the areas between the ceremonial mounds were filled with small residential structures.

away from the high-status burial mounds, with a complete lack of grave goods, indicates that these individuals were commoners and not members of the elite.

Both of these individuals are achondroplastic dwarfs, with full-sized torsos and short, muscular arms and legs. (Achondroplastic dwarfism results from an inherited dominant genetic trait that limits bone growth.) They were both relatively healthy individuals. The male was approximately 50 inches (127 cm) tall, and the female was about 46½ inches (117.9 cm) tall. Both of them exhibit flattening to the backs of their skulls, the common result of being strapped to a cradleboard. This rear flattening served to heighten the bulging of both their foreheads and lower jaws, a common characteristic of achondroplastic dwarfs.

## Mississippian Society

Life for these two people was difficult but it reflects the lives of commoners in Mississippian communities. The teeth of both individuals show extensive wear, the usual result of a diet high in stone-ground maize and its associated grit. Each individual has a large number of caries, with each having lost teeth due to extensive abscesses. The male probably walked with a limp, the result of a hip injury that failed to heal correctly. The female's spine shows that she suffered from arthritis. That both of them lived into middle age, roughly 40 to 45 years, proves that they were functioning members of their social order. The close proximity of the two burials suggests these two dwarf individuals were related – possibly siblings, given the genetic nature of achondroplastic dwarfism.

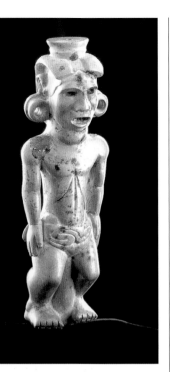

ABOVE An Adena culture soapstone pipe from Ohio, which represents a dwarf. It predates the Moundville burials by over 1,000 years.

Whether they were captives or part of the local population is not known. A normal-sized individual was found buried with the female dwarf, but no records of any analysis connected to this burial appear to exist. The two people may have been a couple, with the female dwarf the spouse of the normal-sized individual. At her time of death, she may have been a later addition to his grave. Even as his spouse, however, it is possible that the female dwarf represents a sacrificial victim included as a burial offering for the normal-sized individual.

Mississippian society was highly stratified into a series of social classes, with the whole ruled by a supreme chieftain. Village rulers and individuals of high social status are often found buried with a wealth of grave goods. Their graves also commonly include sacrificed individuals. These sacrificial victims are usually women and are commonly assumed to have been retainers or captives but could also include spouses. A village chieftain or ruler may have had a dozen people killed and buried with him. Nonruling high-status burials tend to contain fewer sacrificial victims. However, human sacrifices are so rarely found in commoner or low-status burials that it is more likely the female dwarf was a separately buried spouse.

Only one prehistoric representative image of an achondroplastic dwarf is known to exist for North America. This is a tubular soapstone pipe from the Adena culture of the Ohio River Valley. Dating to roughly 1,000 years before the Moundville burials, this pipe is carved in the round in the image of a warrior. He is shown wearing a loincloth, ear-spools and an elaborate headdress. The carved head and torso are to scale, but the figure has the abnormally short muscular arms and legs characteristic of an achondroplastic dwarf. This pipe shows that dwarfs, although perhaps not common, were not unknown.

The excavation of the Moundville dwarf burials reflects the level of scientific thinking of its day. Analysis of these two individuals was conducted, but outside of their cultural context. Questions remain that could still be answered today. Would analysis of the normal-sized individual found with the female dwarf (if the remains could be found) provide insight into their relationship? At present, even its sex is not known. DNA analysis could demonstrate any family relationship between the two dwarfs as well as their relationship to the general Moundville population. Were they captives from another group or part of the local population?

We will probably never know the actual role special individuals, such as dwarfs, played within most prehistoric societies. Were they considered sacred oracles? Were they kept as pets or curiosities? Or were they simply functioning members of their own societies, differing only in their stature or intellect? Burial information of this type plays an important role in the ability of science to understand past cultures. It helps demonstrate the humanity of prehistoric societies by providing insight into their social dynamics.

# Buried with the Friars

**ABOVE** In the 15th century the south aisle of the nave became a chantry chapel, where bodies would be laid in state while services were said or sung for their souls. These services were paid for by relatives, and the body would lie here until that sum of money had been spent. The photo shows a row of four coffin niches, set low down into the side wall of the chapel. To the left can be seen a brick-lined family vault; the buildings to the right are part of the west range of the friary.

From individual skeletons, one can usually assess the stature and age of the person at death; it is also possible to identify the sex of most adult examples with reasonable certainty. Trying to identify diseases is more of a problem, simply because most leave little or no trace in the bones. As this applies to most of the plagues and other diseases documented as having made a dramatic impact on society, this has long been a source of frustration to the archaeologist. Although we can demonstrate that people have suffered from osteoarthritis in most periods, it is far more difficult to identify the impact of a major epidemic such as the Black Death (bubonic plague) or a cholera outbreak.

If we are to be able to say anything sensible about major changes in the health and diet of a community, then we need to look at a large and closely dated sample of bodies. The ideal conditions for such a study can be found in a medieval urban monastery, which fulfilled many of the functions that are now performed by health and social services. Further, they offer a broad cross section of the community – the healthy and the sick, and everyone from beggars to Lord Mayors, and from priests to the laity.

In 1994, the largest such excavation to be mounted in Britain took place at the Augustinian Friary in Hull. Two-thirds of this monastery, which was occupied only from AD 1316 to 1540, was excavated, yielding the remains of over 500 individuals. Not only did the bones survive in very good condition but the waterlogged clays of Hull helped preserve the sort of evidence that often does not survive – the clothing and wooden coffins of many of the burials. Moreover, it proved possible to tie the sequence of superimposed burials to the construction phases of the buildings and to date many of them to within a year through tree-ring dating of the surviving coffins.

## Medieval Burial Tradition

The results have been fascinating and are beginning to challenge some of our assumptions about the origins of certain diseases and about the health of medieval townspeople. Detailed study of these burials has shown that selected parts of the monastery were reserved for particular categories of person. Hence, while the priors were buried exclusively in the chancel of the church, the nave was used for burials of wealthy townspeople. Children and teenagers were buried in the cloister alleys, while the poor, the needy and the ordinary friars were buried in a cemetery to the north of the church. Each of these areas was characterized by its own burial rite – for example, the particular way in which the hands and arms of the skeleton were arranged.

The members of the wealthy lay congregation were buried in their Sunday-best clothes. Textile remains survived on about 20 percent of the skeletons, representing the largest collection of medieval costume fragments ever recovered from a cemetery in Britain. In the later fourteenth and early fifteenth centuries, court fashions changed from brightly colored fabrics to smart new black costumes; several of the burials dating from the period between about 1410 and 1430 display these black fabrics. These costumes would have been worn with newly fashionable canvas breeches; however, one of their drawbacks was that the canvas tended to chafe the thighs, and so, to counteract this, men began to wear underpants. Six of the bodies were found wearing finely woven woolen underpants, among the earliest known boxer shorts. Another man was buried in a full-length gown with a one-piece hood. Others were buried in full-length gowns with voluminous sleeves, fastened at the waist with a leather girdle (a strap that extended down from the belt to the feet). Surviving brooches, buckles and straps on the bodies show how such clothing was fastened in place.

## Bubonic Plague and Syphilis

In 1348 and 1349, the Black Death swept through Britain. This wave of plague is thought to have killed 40 percent of the population. On most archaeological sites, its impact is hard to gauge, but on this one, almost 20 percent of the burials were in wooden coffins made from Baltic oak, the planks of which can be dated by dendrochronology (tree-ring dating) to within a year of the felling of the tree from which they were cut.

**BELOW** The first of a series of burials within a brick-lined vault, which appears to have been a family vault. The body was placed within an oak coffin laid at the base of this vault. By the time of the latest burial, the precise position of the vault was no longer known, and the last burial was placed half in, and half out of the vault.

LEFT The bodies of prominent lay persons from the town were buried in the nave of the church, dressed in their "Sunday-best" clothes. In some cases, textiles have survived; in others (such as this burial), the position of belts is indicated by metal dress fastenings. Here, a copper alloy belt buckle survives next to the pelvis, showing that a belt was worn around the waist.

BELOW Both men and women wore full-length gowns, fastened at the waist. This 14th-century skeleton has a leather girdle or belt around the waist or hips, with a strap that hung down the front of the garment between the legs. Such girdles are illustrated in contemporary manuscripts or depicted on statues of the period; they might be made of braid, or of woven silk, linen or cloth, as well as of leather. This is the most complete example yet found on an excavation.

A cluster of coffins from these plague years probably reflects the onset of the epidemic, with several coffins being made from planks cut from the same tree (i.e., because these coffins were made from the same batch of timber, these burials are likely to have taken place over a very short period). At first the coffins were carefully made, but by the end of the plague years they were being put together like packing crates, perhaps indicating that the carpenters themselves had succumbed to the disease.

After 1390, all of the burials were wrapped in shrouds instead of being buried in coffins. From these later burials comes our first evidence of a new scourge: venereal syphilis. This disease, variously referred to as the Great Pox and the French Disease, is known to have swept through Europe in the early sixteenth century. Its advent was traditionally attributed to the return of sailors from the New World following the voyage of Columbus in 1492. However, at Hull, four skeletons with fully developed tertiary syphilis were present in mid-fifteenth-century levels, showing that the disease was already well established in Europe at least half a century before Columbus set sail. Radiocarbon dates both from this site and from Rivenhall church in Essex provide independent dating to support the stratigraphic evidence (calibrated dates of AD 1435 to 1490 and 1290 to 1445 respectively).

A new research program is planned to look at all of the instances of early syphilis from the British Isles. It will combine radiocarbon dating with DNA analysis (to confirm that these burials did indeed have venereal syphilis) and lead isotope analysis (to identify whether the victims were locally born or were incomers to the communities in which they died) in an effort to develop a clearer picture of the origins of this disease and the pattern of its spread.

Another ailment evident in the Hull cemetery is a bone disorder known to specialists as diffuse idiopathic skeletal hyperostosis (DISH). This is characterized by some of the vertebrae in the spine beginning

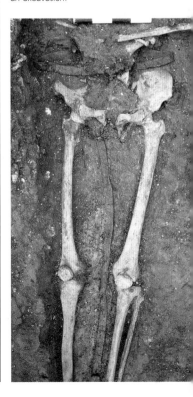

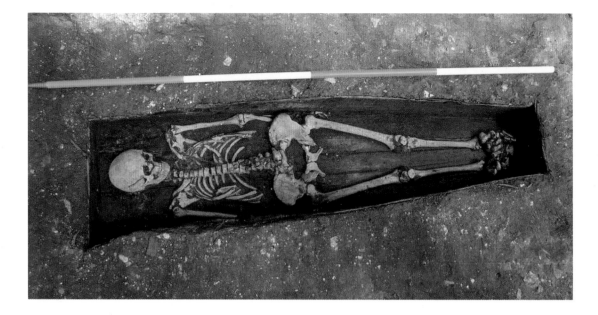

**ABOVE** Many of the mid-14th-century burials were in coffins made of Baltic oak. Here a skeleton can be seen, fully cleaned, lying in a coffin from which the lid has already been removed as part of its excavation.

to fuse together, the extra bone growth being prompted by an over-rich diet, and it is found in overweight, middle-aged men. Previous work suggested that this disorder was more common in closed communities such as monasteries, and its incidence at the Hull friary is some three times higher than would be expected in an average cross section of medieval society. One of the early priors was so fat that he was buried in a heavy oak coffin. By a curious quirk of the soil conditions, his corpulent profile was etched as a red stain on the timber boards at the base of his coffin. With real-life examples like this, is it any wonder that the classic image of friars that has come down to us today is that of Friar Tuck? Though it would be grossly unfair to suggest that most friars were obese, their lifestyle and diet may have assisted this trait.

Several of the burials were accompanied by rods or staves laid either alongside or across the body – some with just one rod, others with a pair. None of these bodies exhibits signs of bone injury to the legs, which might have indicated that these individuals had suffered from lameness and required such support. Pilgrims often carried a staff as a symbol of their status, but none of these burials contains any other regalia that might have denoted a pilgrim. We know that certain monastic officials bore wands of office, but as we have no idea what these looked like, we are not sure whether or not these rods denoted such an official. In one case, the hazel rod is sufficiently slender and of the right dimensions to have served as a flagellation cane.

Lastly, a common practice in the medieval church was to offer prayers for the dead in return for financial donations to pay for the services. This proved to be so lucrative that many churches built so-called chantry chapels to meet the demand and to supplement their income. The Hull friary was no exception, setting aside the south aisle of its nave for this

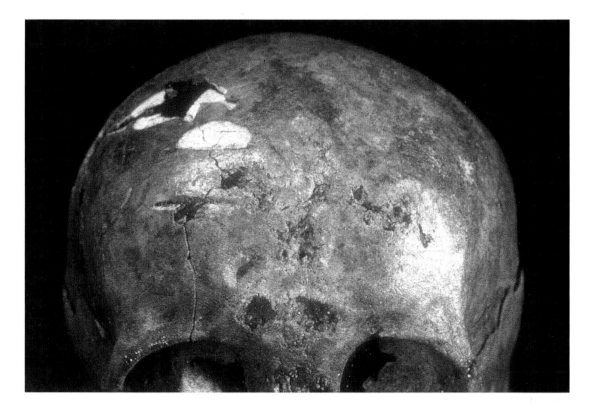

purpose. A row of tomb settings was established in its south wall, where bodies were laid in state for such services; they then were moved for reburial elsewhere when the original donation ran out. A large family vault within this chapel was used so often that the last burial could not be fitted into it; the burial was laid partly in the tomb and partly overhanging its end, suggesting that the ground level in the chapel had been raised so much by the burials that the precise positions of the tomb walls were no longer obvious.

**ABOVE** Skull of a 15th-century skeleton man who suffered from tertiary syphilis. When the disease is this advanced, the sores are not only visible on the skin, but would also begin to eat into the bone. The nose has largely rotted, and the surface of the upper part of the skull is beginning to be pitted with numerous *caries sicca* infections.

**LEFT** Close-up of the surface of the left femur from the same individual, showing irregular compact bone formation caused by the disease.

# Lewis Man:
## A Face from the Past

Cnip headland overlooks Tràigh na Beirgh, one of the loveliest beaches on the island of Lewis in the Western Isles of Scotland. About 3,500 years ago, a man was buried in a cist, or stone-lined grave, in the sandy soil of the headland. A pottery vessel containing food or drink was placed by his head. A low mound of sand and turf was then probably heaped over the grave and edged by a ring of boulders. There his remains lay until a few years ago, when the sharp eyes of local archaeologists Margaret and Ron Curtis spotted a suspiciously regular arrangement of slabs where the ever-present wind had blown away the surrounding sand. As the site lay not far from the remains of a Bronze Age burial cairn excavated during the 1970s, there seemed a good chance that erosion had begun to reveal something new of interest.

Margaret and Ron kept a close eye on the site over the Lewis winter, and the following spring a rescue excavation was organized, funded by Historic Scotland. The grave site was excavated by Andy Dunwell and Tim Neighbour of Edinburgh University's Centre for Field Archaeology. In it, they found the skeleton of the man lying within the remains of the cist and, by the skull, the crushed remains of the vessel. The bones and pot were lifted carefully and removed for study. A sample of bone later provided a radiocarbon date indicating that the burial took place about the middle of the second millennium BC – during the period customarily known as the Bronze Age.

The skeleton was examined by Dr. Margaret Bruce of the University of Aberdeen and Neill Kerr of Aberdeen Royal Infirmary. Their expert analyses showed the remains to be those of a man aged between 35 and 40 years. Even by the standards of the time, he was fairly short, with a height of 5 feet, 4 inches (163 centimeters). But what was of particular interest was evidence of healed but extensive injuries to the right side of his face – in particular, a severe fracture of the cheekbone extending into the orbit of his eye, and also fractures of the jawbone and its joint. The injuries are consistent with a severe blow with a blunt object; one possibility is that the man had been savagely clubbed on the side of the face, but he may have been involved in some kind of accident.

Unluckily for him, infection in the fractured bone ends resulted in complications. Although the fracture of the cheekbone healed cleanly, the mandibular joint healed poorly, without rejoining, and he would have had limited jaw movement, particularly on the right side. This led him to favor a soft diet, resulting in severe dental caries and abscesses in four of his molars and one premolar, in some cases extending deep into the bone of his jaw. He also had bad gum disease, gingivitis, with the deposits of calculus that result from it. His mouth would have given him a great deal of pain, and he probably suffered from halitosis.

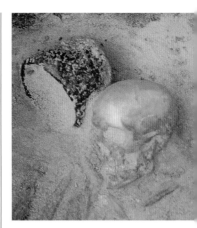

ABOVE The man, aged between 35 and 40 years had been buried with a vessel containing either food or drink.

LEFT View of the grave under excavation.

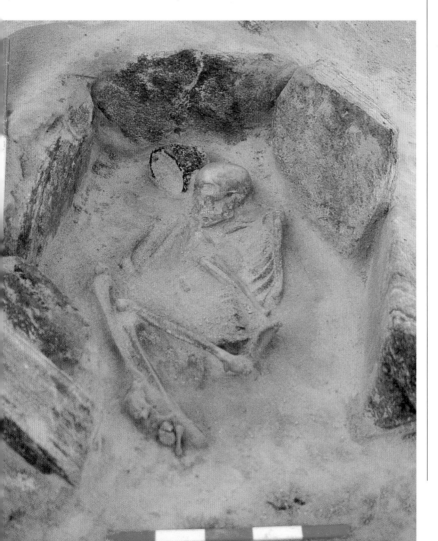

33

## Reconstructing the Face

In view of these interesting features, it was thought that the skull would make an interesting facial reconstruction of the type familiar from police investigations and television documentaries. The aim of the project was to produce a facial likeness by enlisting a team of experts and state-of-the-art medical technology. Supported by grants from the Russell Trust and the Society of Antiquaries of Scotland, work began in the spring of 1995.

The innovative element of the project was to be the production of an exact plastic replica of the skull as the basis for the modeling process. This involved passing the actual skull through the CAT scanner in the Radiology Department of Edinburgh's Western General Hospital. Digital data were then sent to Leeds University, where the plastic replica was manufactured using a then brand-new technique known as *selective laser sintering*.

Finally, the skull and the replica were delivered to Newcastle Dental Hospital, where Brian Hill, then head of the Department of Illustration, modeled the facial reconstruction in terra-cotta. Obviously, certain features – the shape of the ears and nose, cut of hair and length of beard – are simply guesswork. On the other hand, although the wounds had healed, facial surgeons were able to confirm that the man would have been left scarred; the model shows the puckering of the skin of the cheek characteristic of such injuries.

So who was this man? In light of what we know about this period, he was almost certainly a farmer. He would have reared cattle and sheep and grown barley but, like the crofters of the Western Isles in more recent times, his family probably supplemented their diet with the resources of sea and shore, taking fish, shellfish, birds and eggs in season. Deer would have been hunted in the hills, and wild foods gathered.

The Western Isles had been settled for at least 2,000 years. The chances are that he was from the island and that his family had lived there for many generations. We do not know where his farm was, but there is a fair chance that his home lay within sight of the headland where he, and quite possibly a number of his kinfolk, were buried. The landscape in which his farm lay would have looked different from that of today. His ancestors had cleared areas of the woodland that used to cover the islands, to provide grazing for their animals and fields for cultivation, but there were still areas of woodland, particularly on the hills and away from the coast. The blanket of peat that covers the islands nowadays was then only beginning to spread and grow as a result of climatic changes and soil deterioration, perhaps precipitated by factors such as woodland clearance.

ABOVE View of the skull which formed the basis for a reconstruction.

The sea level was lower than it is today, perhaps by about 5 feet (1.5 meters), but slowly rising as it had since the end of the last glaciation. As it rose, it brought with it the shell sands that form the light arable soils of the west coast of the Hebrides, creating a green plain of machair along the shore behind the beach. Through time, settlement became more and more concentrated on this coastal machair, as the growth of peat and exhaustion of the shallow, fragile hill soils pushed farmers away from the inland areas, setting the pattern of coastal occupation that persists in the islands today.

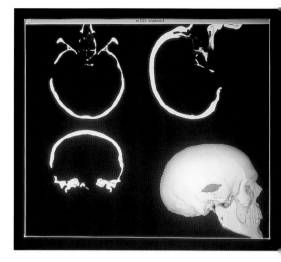

The fact that the man was buried in a specially built cist, close to an established burial cairn, suggests that he was probably a person of some importance. Not everyone seems to have been buried in this way – and it is likely that it was only the top people in the community who received special burial. His age indicates that he was probably a family man and, based on his burial in the cist, probably the head of a family. In many societies, past and present, a man of 40 years of age is likely to be a grandfather, so he may have lived with or near his extended family.

The Bronze Age in the Western Isles was a time of change: social, ecological and economic. It may well be that change brought conflict between groups of islanders with it, particularly as the amount and perhaps the quality of farming land decreased. Being able to lay a claim to coastal arable land would have been important to this man, and the presence of his family burial place on the headland at Cnip may have been as much a part of that claim as his ability to hold, work and defend the farm itself. In truth, we can only guess at how he came by his disfigurement, wondering whether his injuries were the result of a fight or a wider conflict between communities. Perhaps he was simply the victim of an accident in the course of a hard everyday life, for his lower back and left shoulder showed degenerative changes that may have been related to activity as well as age.

The facial reconstruction was created for display along with archaeological material from the Western Isles at the Museum nan Eilean, Stornoway. Of course, many visitors simply flinch at the thought of enduring those injuries and the subsequent discomfort in the absence of modern painkillers and dental care. But it was hoped that the reconstruction would also tempt visitors to think further about the people behind the artifacts in the exhibition. Archaeological evidence is almost invariably silent but, looking at the face, it is perhaps no longer such a leap of the imagination to think about this prehistoric Leòdhasach (inhabitant of Lewis) bemoaning the state of the weather or his crops, casting an eye over his stock or telling stories around the peat fire on a long winter's evening about how much easier life had been in his grandfather's or great-grandfather's time.

ABOVE The basis for the modeling was an exact plastic replica of the skull, manufactured by Leeds University, using a brand new technique, known as selective laser sintering.

BELOW The face of the Bronze Age "Lewisman", as reconstructed by Brian Hill of Newcastle Dental Hospital. The man had once suffered extensive injuries to the right side of his face; his wounds had healed but they would have left him scarred.

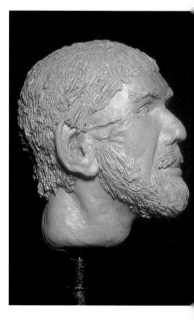

# C18:
## The Pounder from Ur

I n archaeology, only very rarely is a burial found together with the tools of the occupant's working life, other than weapons. Exceptionally, Grave C18, of a mature adult female excavated at Al-'Ubaid in Iraq gives us the opportunity to explore morphological changes to the bones of the skeleton that might have been brought about by the lifelong practice of a specific activity. Naturally, the craft would have had to begin in childhood while the bones were still growing. In the past, as with athletes or musicians of today, work and the acquisition of skills would have started at a very early age.

The skeleton of C18 was recovered from a shallow grave cut into hard soil. By the head was a rough reddishware tankard bowl, a large spindle-whorl, and behind the head a rough stone pounder, almost a cube. The grave is dated to late in the Second Dynasty of Ur, about 2700 BC. The bones were fairly well preserved and constitute the skeleton of a robust female with a strong chin – in fact, at first she was thought to be a male. She also has a pronounced mastoid process and very long styloid process. The wear on her teeth is noticeably flat, as though she had been grinding or gritting them. Her clavicles (collarbones) are enormously expanded, and her breastbone (sternum) unusually broad. The muscle attachment areas on her arms are large, and the articulations of the first metatarsal bones are extended onto the upper surface of the foot.

**ABOVE** Archaeologists initially believed the skeleton to be that of a male, owing to its strong chin.

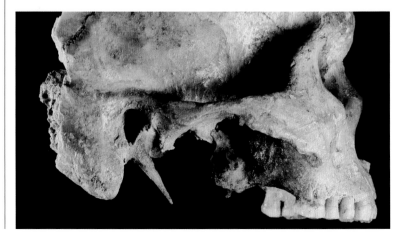

**RIGHT** Closer examination of the teeth revealed that the female must have grinded or gritted them repeatedly. There was considerable wear, and they were flat.

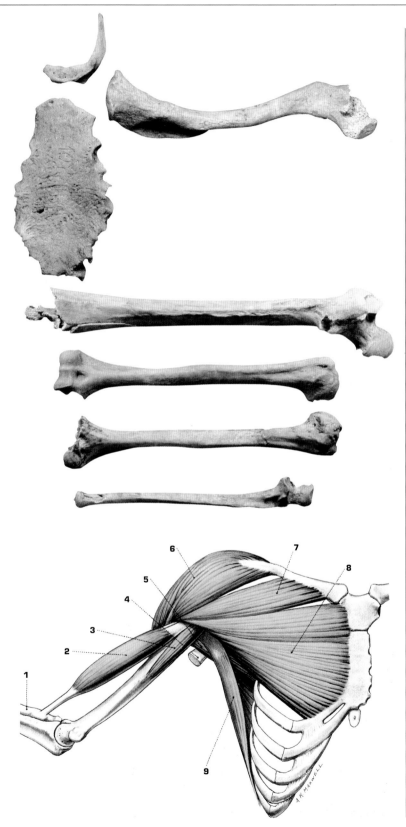

LEFT C18's hyoid, collarbone and breastbone were unusually broad when compared to those from other skeletons.

LEFT The arm bones have strongly developed areas where the muscles were attached. The thigh bone (below) has strongly developed attachment areas for muscles that pull the thighs together.

LEFT Bones and muscles of the upper limb.

1 *Radius*

2 *Biceps brachii*

3 *Coracobrachialis*

4 Long Head of *Triceps brachii*

5 *Teres major*

6 *Deltoid*

7 Clavicular Head of *Pectoralis major*

8 Sternocostal Head of *Pectoralis major*

9 *Latissimus dorsi*

## What Does All This Mean?

The teeth have flat wear, which suggests that she habitually gritted them. The large mastoid indicates that there were strong mastoid muscles, which insert on the medial end of the clavicle. If the muscles on both sides contract, the head is flexed, but if the head is fixed, the muscles elevate the sternum and first rib. Strong development of the insertion areas for the jaw's muscles give it a masculine appearance. These muscles are involved in closing and opening the jaws. The styloid is unusually long and well developed for a female, and the hyoid (or tongue bone) is robust. Muscles between the hyoid and styloid are involved in stabilizing the jaw, in a fixed position, when the weight of the body is required to exercise force on an outside object.

The clavicles both have a remarkably broad area for the attachment of the clavicular part of *Pectoralis major*, which must have been a massive muscle. The clavicular head of *P. major* flexes the upper limb at the shoulder and the sterno-cleido head extends the flexed limb against resistance, as when downward pressure is exerted on an external object. *P. major* also inserts onto the sternum, which is greatly expanded, and onto the humerus, which is robust. The *P. major* muscle is essential in activities that require stabilization of the shoulder when the arm and the body are held close together. Strength in these muscles is

**ABOVE** During the action of pounding – using the weight of the body to exercise force on an outside object – muscles between the hyoid and styloid hold the jaw in a fixed position.

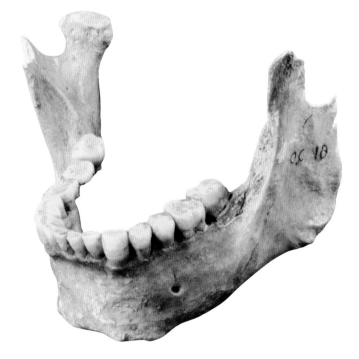

**RIGHT** Strong development of the muscles involved in opening and closing the jaw led to its masculine form.

necessary for activities where the body weight is taken through the upper limbs and the torso moves forward and backward. As a whole, *P. major* takes an active part in the movements of adduction and medial rotation of the humerus, but the activity is only marked if resistance is to be overcome – as when using a pounder?

The forearm bones have strongly developed areas where the muscles that turn the arm are attached. Thanks to these muscles, objects that may be heavy can be picked up against gravity with the forearm turned. The thigh bones (femurs) also have strongly developed ridges for the muscles that pull the legs together. The first metatarsals of the foot have an extended proximal articulation that is associated with a habit of kneeling with toes curled under.

This combination of traits points to a powerful activity involving forced pressure of arms, with forearms turned. You grit your teeth when using a screwdriver so that you can obtain a greater force downward. This uses the strength of the chest muscles, especially *P. major*.

Of course, *P. major* is also developed in such activities as rowing and using a crutch. (The woman from Jericho on display in Gallery 56 of the British Museum was buried with her crutch. She too has an expanded sternum.) C18 was found with a large pounder in her grave. The changes are consistent with a lifetime spent grinding, possibly ocher for dyes. Being buried with the tools of her trade takes some of the guesswork out of understanding the bones of C18.

So we have a picture of this strong woman energetically, and with teeth clenched, working away with the heavy pounder that was found in her grave, rather like the women photographed in Africa pounding millet.

**ABOVE** Female pounders from former northern Rhodesia. They have powerful arm muscles; their thigh muscles are pulled together and the toes of the woman on the right are curled under.

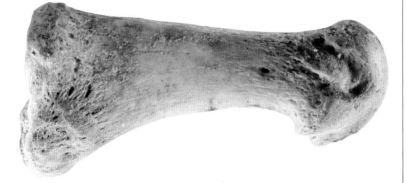

**LEFT** Extended joint surface of her first metatarsal allowed C18 to kneel with her toes curled under.

# The Einiqua:
## People of a Hot, Dry Frontier Land

RIGHT A sketch of the burial cairn of a Khoekhoe chief near Kakamas drawn in 1779 by Colonel R. J. Gordon, who probably witnessed Einiqua burials. The cattle and sheep bones, as well as the entire ox head, indicate a wealthy or important individual.

*Hottentots Capiteins, Graf.*

T he western interior plateau of South Africa is a sun-baked, arid land through which the mighty Orange River, 1,243 miles (2,000 kilometers) long, cuts a green, Nile-like swathe from the mountains of Lesotho to the Atlantic Ocean. Known as the !Garib, meaning "wilderness" or "desert," to local precolonial peoples, its lush banks provide a stark contrast to the barren, rocky scrublands of Bushmanland to the south and the ocher-colored sands and thorn trees of the southern Kalahari Desert to the north. In its middle reaches, a short distance downstream from the agricultural town of Kakamas, the river boils through the spectacular 5.6-mile (9-kilometer) -long gorge of the Aughrabies Falls, the "place of great noise."

RIGHT Rarely preserved soft calcified tissue found with this skeleton of a 50- to 60-year-old woman, excavated by Alan Morris from a grave below a cairn at Omdraai near Kakamas, suggests she died of kidney disease.

In the eighteenth century, this area was a vibrant place of contact between four communities. There were small populations of indigenous hunter-gatherers, known historically as San or Bushmen, whose stone tools indicate that their ancestors lived there sporadically for hundreds of thousands of years. There were also indigenous Stone Age pastoral people with pottery, domesticated sheep, goats and cattle. These people are known collectively as the Khoekhoen (also Khoikhoi); they lived in territorial clans and moved about in search of grazing for their livestock. Although it is not known how long they had been there, the Khoekhoen were present elsewhere in western South Africa after about 2,000 years ago. Third, in recent centuries, the summer rainfall areas to the east were occupied by settlements of black Iron Age farmers whose ancestors migrated from farther north to southern Africa in various stages over the past 2,000 years or so. Finally, there were Europeans from the Cape Colony, which had its beginnings as a refreshment station established by the Dutch East India Company at the Cape of Good Hope in 1652. The Europeans wished to acquire land and establish trading relationships with the locals on their northern frontier.

The first historical description of the people of the Aughrabies Falls was made by Hendrik Jacob Wikar, who recorded his travels in the region in September 1778 and April 1779 in a journal accompanied by a sketch map. He referred to the Khoekhoe inhabitants of the region as the Eynikkoa or Einicqua. A few months later, the accomplished naturalist Robert Jacob Gordon visited the area and identified people he met at the Falls as Einiqua.

The publication of Wikar's journal in 1935 inspired two South African zoologists, T. F. Dreyer and A. J. D. Meiring, to find the graves of the Einiqua in the hope of making a collection of skulls of undoubted

LEFT The excavation of the Omdraai skeleton in progress: the large flat stone in the foreground was exposed by the partial removal of the cairn and capped a grave shaft in which the remains were found.

Khoekhoe origins. The following year, they excavated 82 graves and kept 56 skeletons, now housed in the National Museum in Bloemfontein, South Africa. To supplement Dreyer and Meiring's poor documentation, A. G. Morris excavated five more skeletons near Kakamas in 1984. These are housed in the Department of Human Biology at the University of Cape Town. The Einiqua skeletons constitute one of the largest collections of precolonial human remains from South Africa and provide remarkable insights into the health, lifestyle and relationships with contemporaries of indigenous South African pastoralists in the period just before European colonization.

## Dating the Skeletons

All living things contain minuscule amounts of carbon 14, a radioactive form of carbon that decays at a known rate and is replaced in living organisms through exchange with the atmosphere. However, when the organism dies, the carbon 14 that decays cannot be replaced and so gradually diminishes until none is left. After about 40,000 years have passed, so little is left that it is difficult to detect. Measurement of the carbon content of remains like bone and charcoal younger than about 40,000 years can indicate how much time has passed since death.

Radiocarbon dating of bone samples from the Einiqua skeletons indicated that they died in the middle to late eighteenth century and are very likely from the groups recorded by Wikar and Gordon. Interestingly, however, one set of imported glass trade beads found with a skeleton was of a kind made between the sixteenth and nineteenth centuries, another was of fifteenth- to eighteenth-century origin, while a third was at least pre-sixteenth century, possibly as old as the tenth to thirteenth centuries. Some southern African societies keep beads as heirlooms, a practice that would explain the discrepancy between the bead dates and the radiocarbon dates, and also suggests that the graves in which they were found could have been of individuals of power and prestige.

## Health and Lifestyle

Wear on the chewing surfaces of teeth is an excellent indicator of the kind of diet and food preparation methods used by people, while the number of cavities in the teeth and the number lost before death reflect both diet and dental hygiene. Traditional hunter-gatherers, whose staple diet consists mainly of wild plant foods with some meat and very little sugar or carbohydrates, have worn teeth but a low incidence of cavities.

African farmers who live in rural villages and subsist on a diet of cereal porridge with sour milk and no sugar have less wear on their teeth but more cavities, probably because the porridge sticks between their teeth and provides a haven for bacteria. City dwellers who eat softer food and more sugar have considerably less wear on their teeth but far higher rates of cavities.

The teeth of the Einiqua are worn but indicate that they rarely suffered cavities. Although the Einiqua were pastoralists, their dental records suggest a staple diet similar to that of traditional hunter-gatherers. This is corroborated by early historical accounts, such as that of Petrus Borchardus Borcherds, who in 1801 recorded that the common food of the Khoekhoen consisted of honey, locusts, gum and wild plants, while they were "very sparing" of their cattle and sheep.

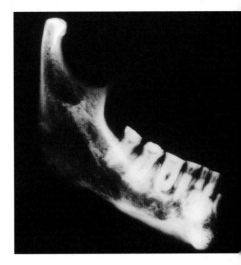

**ABOVE** This radiograph of the lower jaw of an adult Einiqua female shows a benign dental tumor – a cementoblastoma – at the root of the left canine tooth.

## Joints

Twenty percent of the adult skeletons have evidence of arthritis, an inflammation of the joints that results in the destruction of the joint surface and abnormal growth of bone around the edges. All but one of the Einiqua cases are probably the result of degenerative joint disease associated with aging and a physically strenuous lifestyle. Unlike modern urban people, whose knee and hip joints are most commonly affected by arthritis, in the Einiqua the joints most commonly affected are in the shoulder, elbow and back. This suggests they used their arms and thorax more than people today. One case of arthritis was probably caused by infection. This occurred in a young adult male whose bones of the lower arm, wrist and hand fused, probably as a result of a deep wound that became septic after a fracture of the lower arm.

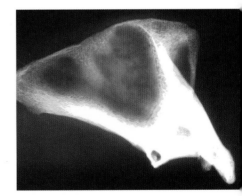

**ABOVE** Radiograph of a shoulder blade of an Einiqua individual, showing a soft-tissue tumor just below the shoulder joint.

## Eye Sockets

Porous bone growth in the roof of the eye socket, known as *cribra orbitalia*, is probably a reflection of iron deficiency. Three Einiqua children, who were 18 months, 8 years and 12 or 13 years old respectively when they died, suffered from this. However, this incidence is low, and it seems the Einiqua were comparatively well-nourished.

## Individual Cases

Two adults show evidence of mild cases of spina bifida, a congenital defect resulting in incomplete formation of backbones; one adult has spondylolysis, a separation of the arch of the vertebra into two parts, which would have caused chronic lower back pain and more serious difficulties if slippage occurred; and an elderly female has scoliosis, or a lateral curvature of the spine. Two adults have bone tumors, one a benign mass at the root of a tooth and the other an enlarged marrow cavity and thin outer layer of a lower arm bone that points to the existence of a tumor. In addition to the young adult male with infective arthritis probably caused by a fracture, there is also an old adult female with a clearly defined healed break on her nasal and upper jaw bones, probably caused by a sharp object like a knife. One of the most interesting cases is the recovery of soft calcified tissue from the skeleton of an elderly female, a rare event in archaeological situations. Changes in the structure of her hip bone indicate she was about 50 to 60 years old at death. During excavation, two calcareous deposits were recovered from the place where her kidneys would have been located. Investigation of the masses by X-ray and scanning electron microscopy identified the material as apatite, which is commonly found in kidney stones. The size and shape of the masses suggests that the female suffered from severe kidney disease and that renal failure was almost certainly the cause of her death. She would have been very ill and would have had to be cared for by others for months before she died.

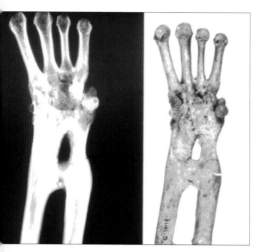

**ABOVE** The damaged right hand of an adult Einiqua male (radiograph on the left) showing that the bones have fused together as a result of trauma to the carpal region (the heel of the hand).

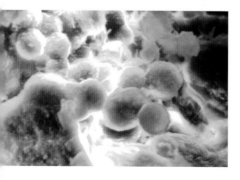

**BELOW** Examination of microscope images like this, at 3,600 times magnification of the calcified material found with the adult Einiqua female from Omdraai, suggested that she probably died from kidney failure.

## Who Were the Einiqua?

Dreyer and Meiring were convinced they had found skulls of "pure" Khoekhoen. Certainly the graves, which were marked with a large stone cairn and a rarity of grave goods, are typical of those associated historically with Khoekhoen. However, Morris's more detailed study of some 60 measurements on each skull and comparisons with those from other population groups showed that they have a mixture of Negroid features and those characteristic of traditional southern African hunter-gatherers and pastoralists. Historical records confirm that gene flow through marriages occurred between the Khoekhoe peoples of the Orange River and their black neighbors, some of whom were essentially pastoralists too. In fact, the homogeneity of the Einiqua collection suggests that the gene flow had been going on for a long time. Clearly, the delineation of groups on the northern frontier and relationships between them were more complex than apparent differences in appearance, economy and culture led early investigators to believe.

## CHAPTER TWO

# natural deaths

C ause of death will always remain unknown for the vast majority of ancient humans, whose remains come down to us as skeletons or cremations, since most fatal diseases only affect soft tissues. However, in a few exceptional circumstances – where bodies are discovered in a fairly intact form – autopsies can be carried out to pinpoint the precise cause of death.

# Bodies from the Ashes:
## Herculaneum and Pompeii

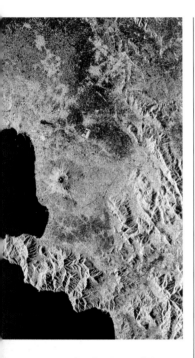

## The Ancient Accounts

The eruption of Vesuvius on August 24, AD 79, engulfed the prosperous cities and villas round the Bay of Naples. Among the victims was Pliny the Elder, author of *The Natural History*, who was in command of the fleet at Misenum, 10½ miles (17 kilometers) to the west of Pompeii. He had attempted to rescue those who had been caught up in the disaster and was overcome by fumes when he went ashore near the town of Stabiae. Pliny's nephew, Pliny the Younger (a future governor of the Roman province of Bithynia), wrote a detailed description of his uncle's fate to the Roman historian Tacitus. The analytical ancient account can be interpreted by modern vulcanologists to explain the series of events that led to the serious loss of life in cities like Pompeii and Herculaneum.

## The Excavations and the Discovery of Bodies

The buried cities lying below Vesuvius were, in effect, lost until the eighteenth century. The taste for classical art was a driving force behind the excavations, with sculpture and wall paintings being removed for display in the royal court at Naples. The haphazard exploration of Pompeii was transformed into a scientific excavation by Giuseppe Fiorelli, who worked on the site from 1863 to 1875. He noticed hollow spaces in the ash and realized that by filling them with plaster of paris he could obtain a cast. This technique, first used in 1870, provided the excavators with a glimpse of the final moments of the citizens of the city as they were overcome by the disaster.

## Prelude to the Disaster

The whole region around the Bay of Naples is affected by volcanic activity and earth tremors. On the north side at Pozzuoli were the Phlegrian Fields, part of the caldera. One of the warning signs of the disastrous eruption of AD 79 was a major earthquake that hit the area around the Bay of Naples on February 5, AD 62. One of the last major eruptions had been more than 1,000 years before; the geographer Strabo, writing under the first Roman emperor Augustus, had noted the inactivity of the mountain and the fertility of the surrounding region.

Excavators have found traces of buildings damaged and under repair when the major eruption engulfed Pompeii. A vivid image of the effects of the earthquake of AD 62 was carved on a relief in a small household shrine (*lararium*) in the house of the banker Lucius Caecilius Jucundus; it shows the collapse of the temple of Jupiter and the damage to the equestrian statues that stood on podia flanking the main frontal flight of steps. The temple, lying at the northern end of the civic forum, was still under repair in AD 79 and may even have served as a temporary sculptor's workshop. In subsequent years, the magma chamber under Vesuvius swelled, causing the land to rise and the sea to recede.

## The Sequence of the Eruption

The description of the younger Pliny allows the several elements of the eruption to be followed. Assumptions must be made about the precise timing of the events, but the rate of fall of ash can be taken from Boscoreale, the site of a sumptuous villa between Pompeii and Vesuvius. The first eruption and subsequent ash fall probably took place in the early hours of August 24, AD 79. Excavations at the rural villa at Terzigno, only 3.7 miles (6 kilometers) east of the crater, show that the deposits of ash had not been disturbed before the major falls later in the day. The eruption was first observed from Misenum at around 1:00 pm on the same day.

A cloud "like a pine tree" was seen emerging from the summit of Vesuvius. This cloud, reaching an estimated 16.8 miles (27 kilometers), was formed by the ejection of a mass of ash and pumice from the volcano. From the deposition of ash layers, it is clear that the cloud moved in a southerly direction. The vulcanologist Haraldur Sigurdsson was able to study the deposits and link them to Pliny's sequence.

At this point of the eruption, Pompeii started to be covered in a layer of ash and pumice falling from the cloud. Estimates suggest that this happened at a rate of some 6 inches (15 centimeters) an hour. The elder Pliny experienced this phase as he arrived near Stabiae: "Ashes were already falling, hotter and thicker as the ships drew near, followed by bits of pumice and blackened stones, charred and cracked by the flames" (Pliny the Younger, Letters 6.16 [trans. B. Radice]). This rain of lapilli, up to 2 inches (5 centimeters) in diameter, was damaging, and Pliny and his companions were forced to tie pillows to their heads for protection. The force of the debris was probably enough to kill.

The area under the cloud was by now cut out from the light; Pliny the Younger recorded that the "darkness [was] blacker and denser than any ordinary night" (Letters 6.16 [trans. B. Radice]). By the evening,

ABOVE The forum of Pompeii lined with colonnades. Vesuvius can be seen behind the temple of Jupiter. The forum had been damaged in the earthquake of AD 62 and was undergoing repair at the time of the eruption in AD 79.

ABOVE The *Eruption of Vesuvius* by 18th-century Italian artist Carlo Bonavia.

approximately 4 feet, 3 inches (1.3 meters) of ash and white pumice accumulated at Pompeii, but the eruption continued for several more hours. At this stage, the weight of the ash on the roofs led to a series of collapses, and people were no longer able to hide indoors. Another layer of gray pumice, more than a meter thick, fell, so that by the early hours of the morning up to 8 feet, 2 inches (2.5 meters) of debris had been deposited over Pompeii and nearly 5 feet (1.5 meters) at Boscoreale.

At this point, the eruption changed form. The mass of rock pushed upward from the volcano started to come down on the slopes of Vesuvius. This initiated what vulcanologists call *pyroclastic surges* and *pyroclastic flows*. This phenomenon, sometimes called *Peléan* after an observed eruption of Mount Pelée on Martinique in 1902, involves a mass of hot gases that surge down the sides of the mountain at speeds of up to 62 miles (100 kilometers) per hour. The flows are denser and tend to follow natural features, like valleys. Evidence for both flows and surges has been found in the deposits of the cities and villas left by the eruption of AD 79.

The first pyroclastic surge seems to have hit Herculaneum around 1:00 am on August 25. It took perhaps four minutes to reach the city. The force toppled colonnades and moved building materials up to 13 feet (4 meters). People were killed instantly as they choked in the ash-laden atmosphere. This surge was followed by a flow that swept round the city and covered the beach.

The third pyroclastic surge, at around 6:30 am, seems to have reached Pompeii itself; evidence has been found by the Herculaneum gate at Pompeii. Structures outside the walls, such as the Villa of the Mysteries and the Villa of Diomedes, were destroyed at this point. Herculaneum was engulfed by the subsequent pyroclastic flow, which covered all but the theater and pushed back the coastline. Initially, Pompeii was protected by its walls, but a fourth surge, at around 7:30 am, broke over the top, assisted by the buildup of ash, and engulfed anybody left in the city.

It has been observed that many of the bodies found at Pompeii were recovered from the top of the layer of the pumice, between the layers linked to the fourth and fifth surges. Apparently these people had abandoned their houses but in the darkness had been unsure in which direction to flee. The wave of hot gases killed them instantaneously. Some estimates suggest that about 10 percent of the population of Pompeii was killed at this point. Traces of the fourth surge have been found over ½ mile (1 kilometer) to the south of Pompeii and nearly 2 miles (3 kilometers) to the east.

The sixth pyroclastic surge took place around 8:00 am. Pliny the Younger described it as a "fearful black cloud – rent by forked and quivering bursts of flame, and parted to reveal great tongues of fire, like flashes of lightning magnified in size" (Letters 6.20 [trans. B. Radice]). They watched it sinking and spreading over the Bay of Naples. Along with other residents of Misenum, the younger Pliny and his mother were forced to flee, even though they were 20 miles (32 kilometers) away from the eruption. Such an extensive surge may have killed people, including the elder Pliny, who had managed to flee the cities close to Vesuvius.

## Bodies from Herculaneum

Herculaneum, unlike Pompeii, was not engulfed in ash, and the bodies could not be extracted by Fiorelli's techniques. It was even suggested that most of the population of Herculaneum had escaped, as so few bodies had been found; 10 were discovered prior to 1982. These included a baby, still in its cot, in the House of the Mosaic Atrium, and two bodies in the baths.

This view was challenged when, in 1982, excavations along the line of the ancient seashore at Herculaneum brought to light a number of new bodies in what appear to have been boat sheds. The moisture seeping through the layers had helped preserve the bones from decay. Some of the chambers contained up to 40 bodies. In one was what has been interpreted as a family. There were three men (aged 35, 31 and 25), four women (42, 38,

**BELOW** Giuseppe Fiorelli found that he could fill gaps in the ash at Pompeii with plaster of Paris. This allowed him to recover a cast of human bodies, animals and other organic material.

16 and 14), and five children (10, nine, five, and three years old, and seven months). The 14-year-old woman had been subject to particularly heavy labor, identified by scars on the humeri, and her teeth indicated that she had suffered a major illness when aged 11. She was found cradling the baby, and it is possible that she was a household slave.

The context showed that these two had been killed by the first surge, which hit at around 1:00 am. The skeletons were not carbonized, suggesting that the temperature of the surge was not extremely hot. However, some parts of the bodies that protruded above the initial deposit were carbonized when the second surge hit the town around an hour later.

Most of the bodies in this beach area were found oriented toward the chambers, and most of the victims died lying on their fronts or their sides. Comparisons with those killed after the eruption of Mount St Helens in 1982 suggests that the most likely cause of death was the airways being blocked with ash.

A study of the bodies allows further details to be extracted. A Roman soldier was identified by the sword at his side. He had probably been seconded to the region on a project, as he was carrying a set of carpentry tools. Analysis of the bones and teeth suggests that he was 37 years old. There were traces of a possible stab wound on his left femur, and other bone wear suggests that he was used to riding horses.

A blond-haired woman, aged about 25, was found to be pregnant with her first baby. Another woman, aged around 35, had particularly strong upper limbs, suggesting to archaeologists that she was a weaver. The force of the pyroclastic surge was indicated by the body of a woman lying on top of broken tiles; one of her thigh bones had been thrust through her body into the clavicle. She was probably hit by flying debris.

The layers of deposits are important for understanding the sequence of events. A body was found beside a boat some 26 feet (8 meters) long, lying keel up. Initially, it was thought that the person was the helmsman. However, it became clear that he had been killed during the first pyroclastic surge and that the boat had been pushed alongside him when the flow swept along the beach.

The first pyroclastic surge that rushed down on Herculaneum is likely to have taken more victims than previously allowed. The residents of the town may not have realized that they were in so much danger, as less than ½ in (1 cm) of ash had been deposited during the first phase of the eruption. It now seems that hundreds and possibly thousands of people were killed on the beach in the early hours of August 25, AD 79.

**BELOW** The body of a 45-year-old woman excavated in a probable boatshed on the beach at Herculaneum. Gold rings are still on her fingers and next to her was found other gold jewelry.

# The Wife of the Marquis of Dai

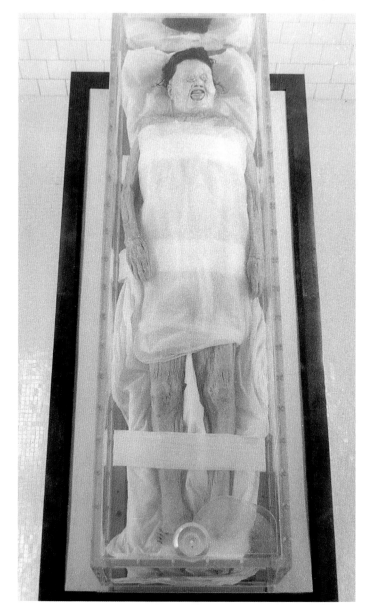

C hinese scholars had long suspected that two small hills in the eastern part of Changsha city hid ancient graves, as had been reported in dynastic history. But it was not until 1971 that archaeologists in Hunan province hit the jackpot. Construction workers digging the foundations for a new hospital accidentally brought to light one of the greatest discoveries in the history of Chinese archaeology: three tombs of the family of the Marquis of Dai.

Scientific excavation started in April 1972 with Tomb 1. The excavators discovered that tomb robbers had made two holes, but they had failed in their attempt to reach the wooden chamber. The tomb contained the remarkably well-preserved corpse of a lady who had died in her fifties, and more than 1,000 grave goods. Tombs 2 and 3 were discovered in 1973, making possible the identification and the exact dating of all three graves through the inscriptions on seals and bamboo strips. Tomb 2 belonged to Li Cang, prime minister of the king of Changsha, who received tribute from the little state of Dai. He died in 186 BC. His son, buried in Tomb 3, died almost 20 years later (168 BC) at the age of 30. Tomb 1 is clearly of later date than the others because it was built on top and caused damage to tombs 2 and 3 as a result. Tomb 1 remained intact and untouched and was attributed to Xin Zhui, Lady Dai, the wife of Li Cang and the mother of the owner of Tomb 3.

The mummy of Lady Dai's corpse was placed in an extended position in the innermost coffin, with her head to the north. The corpse itself was

51

5 feet, 1 inch (1.54 meters) tall and weighed 76 pounds (34.5 kilograms). The body was wrapped in 20 layers of garments and bedding, all of which was finally tied up horizontally from head to foot with nine bands. A floss-wadded robe of yellow gauze and deep red silk embroidered with a *chang shou* (longevity) design was placed on top.

Besides the light plain or painted gauze dress and silk gowns lined and hemmed with gauze, there were woven silk dresses ornamented with embroidery. Silk fabrics and garments of such quantity, high quality and variety are exceedingly rare in Han Dynasty tombs of this kind. At the time of the discovery, all the garments could be unfolded and were still in very good condition.

Lady Dai's hands held two incense bags and were bound together with her feet by a hemp cord. She wore silk socks and shoes with a double-pointed and upward-curving tip. Three hairpins and 29 ornaments held her black wig of natural hair. Her eyes and face were covered with two pieces of fine silk. Her eyeballs had fallen out, but her light yellow complexion and some eyelashes were clearly visible. The eardrum was still intact. Her mouth was wide open, and her tongue protruded a little. Sixteen teeth were left in her mouth, where a small bag with uncooked rice was placed in conformity with the ancient rites for correct burial. A cylindrical pillow placed directly under the jawbone supported her chin.

The autopsy of the corpse started in 1972. Beforehand, a preservative was injected into the corpse, and the muscles gradually absorbed it. The yellowish-brown body was completely intact and moist, while the soft tissues, including skin and muscles, were elastic. In addition, the joints were still flexible. The inner organs had a certain elasticity, and their microstructure was in an excellent state of preservation. Some dislocations visible in the lower part of

**BELOW** When opening up the innermost coffin, archaeologists found the body of Lady Dai wrapped in silk clothes and tied up with nine silk cords.

the body may have been the result of an early stage of decomposition caused by bacteria in the stomach. Once the tomb was closed, the lack of oxygen stopped this process. The skeleton was complete, with all bones in their correct position; they showed traces of osteoporosis.

Through X-ray examination, the small nasal bone was also visible. All the inner organs were intact in both conformation and microstructure. Solid blood – identified as type A – remained in the arteries. The results of the autopsy showed that Lady Dai suffered from at least 10 diseases, but there was no emaciation or indication that she had been confined to bed for a long time. In fact, none of the diseases, including coronary arteriosclerosis, cholelithiasis (gallstones), a distinct narrowing of the fourth lumbar space and pulmonary tuberculosis, had led directly to her death. Instead, 138 muskmelon seeds found in her stomach indicated that this overeating had caused a complete breakdown of her body system (complicated by the cholelithiasis), leading to a sudden heart attack.

**ABOVE** Food, including meat, fish and a variety of fruit and vegetables were placed in bamboo baskets tied up by cord bindings. The contents were then listed on small wooden labels attached to the baskets.

## Preservation

The funeral rites for aristocrats, as described in ancient Chinese books like the *Liji* (*Book of the Rites*), were exquisite and highly sophisticated. When a relative of the nobility died, the body was first washed in a fragrant spirit (called *chang*), then wrapped in cloths and garments. Before the burial ceremony, the corpse was laid on a wooden plank and placed on ice or in cold water. It was then laid in the innermost coffin of several and an acid liquid poured in. This liquid stopped the growth of enzymes and functioned as a mild antiseptic. When Chinese archaeologists first opened Lady Dai's innermost coffin, they found the corpse lying in a red liquid at a depth of 8 to 12 inches (20 to 30 centimeters). No exact analysis of the liquid has been possible to date, but written sources from later dynasties tell of conservation methods involving cinnabar. Only three years after the excavation of the Mawangdui tombs, the discovery of a large tomb in Hubei province, Fenghuangshan district, shed new light on this unresolved question (see Fenghuangshan Tomb 168, p. 59).

## The Tomb Chamber

In Lady Dai's tomb, four coffins were perfectly nested, so little air remained within. The rest of the oxygen was consumed immediately after closure by the oxidation of the organic material. The famous T-shaped banner showing the heavenly realm of the deceased lady, used during the funeral procession, covered the innermost coffin.

The coffins were then placed in a large tomb chamber. The chamber comprised a central compartment for the coffin and four side rooms for the grave goods. Astonishingly, no metal nails were used for this construction; the logs were fitted together by mortises and tenons or wooden nails. For the construction of this chamber, the carpenters used 72 cypress wood planks (logs) – a total of 57 cubic yards (43.6 cubic meters) of timber. The wooden chamber was 22 feet (6.73 meters) long, 16 feet (4.81 meters) wide, and 9 feet, 2 inches (2.8 meters) high, and covered by 26 bamboo mats. It was placed on a thick layer of sand covered with white clay to protect the wood from groundwater. More than 10,000 pieces of charcoal formed a tight layer, 12 to 20 inches (30 to 50 centimeters) thick, that directly enveloped the chamber. Next, a layer of white clay, 3 feet, 3 inches to 4 feet, 3 inches (1 to 1.3 meters) thick, was placed in the shaft, which was finally filled with earth.

Changsha's red soil is acidic, which is normally not favorable for

**ABOVE** The northern compartment was furnished in the style of a reception room. The walls were embellished with silk curtains, and food and drinking vessels were placed in front of the pillow, on which Lady Dai was supposed to sit.

the preservation of either wood or bone. Thus, the thick white clay and the isolating charcoal were responsible for the tomb's perfect state of preservation; they formed an airtight seal and allowed no water penetration. A constant level of humidity and temperature was maintained in the chamber.

## Food and Medicine

Textiles, a bronze mirror and 69 peachwood figures were placed directly in the coffin; all the other grave goods were laid out in the four side compartments. The inventory, listed on 312 bamboo slips, included the objects (two-thirds of which related to food and drink) that were "intended to be buried" but was not identical to the actual arrangement.

The northern compartment was laid out as an inner chamber with curtains, a bamboo mat, eating and drinking vessels, and wooden servant figurines. The western compartment housed cosmetic articles, mostly lacquered wooden objects, and furniture; the eastern one contained figurines clothed as dancers, musicians and attendants. Forty-eight bamboo storage baskets were found intact. These plaited and lidded containers were tied up with hemp cords; some had clay seals and a wooden label indicating their contents or the person in charge (e.g., "Major domo of the household of Marquis Dai"). Six contained garments and silk fabrics, 30 held foodstuffs, three held herbal medicines and four held coins and elephant teeth.

All of the food and medicine has been identified. Agricultural products like rice, wheat, oats, millet, soybeans, red lentils, hemp seeds, plums, pears, arbutus, jujube, mustard greens, ginger and lotus root were placed in bamboo cases or hemp sacks, most of them perfectly preserved and sealed with a small clay seal. The inscription on the wooden label identified the contents. Scientific research, especially on the fruiting season of the plums, proved that the tomb probably had been closed at the beginning of summer. Moreover, the discovery of animal bones belonging to ox, sheep, pig, deer, dog, hare, crane, wild goose, duck, pheasant, chicken, sparrow and carp showed the variety of meat products used for food in the Han Dynasty.

The earliest extant specimens of Chinese medicinal herbs had been placed in bags or hemp sacks, perfume satchels, braziers and pillows. The medicine books found in Tomb 3 record over 240 kinds of medicine: orchid, magnolia, fragrant reed, cassia ash, wild ginger, etc. Some of these were also used to purify the air.

Beside philosophical texts, records of 52 diseases were discovered on bamboo strips, with 283 prescriptions concerning various branches of Chinese medicine – internal medicine, surgery, pediatrics, and the five organs – as well as records of local diseases. These records serve as important examples of early Chinese medical treatment.

**ABOVE** Wooden figurines placed in the northern compartment of the tomb chamber wear clothes showing contemporary fashion.

**BELOW** Precious lacquer objects, including a tray with plates, drinking cups and bowls, demonstrated Han Dynasty luxury. Lacquer objects were so highly valued that they cost more than bronze vessels.

# The Triple Burial of
## Dolní Vestonice

S ince they were first discovered and authenticated in the late nineteenth century, Upper Paleolithic burials – that is, those dating between about 30,000 and 10,000 BC – have become quite numerous, although still rare in the context of the period's length and the number of people who must have died. It seems likely that many of those accorded careful or elaborate burial in the last Ice Age were special (see pages 21, 108 and 148).

This was certainly the case with a unique triple burial that was excavated by the late Bohuslav Klíma at the famous open-air site of Dolní Vestonice in the Pavlovské Kopce Hills of southern Moravia, where decades of work have provided many insights into Early Upper Paleolithic life – not just camps and burials but also remarkable terra-cotta figurines and their kiln.

Whatever the events and medical history behind this triple grave, it adds to our picture of the complex nature of beliefs and funerary rituals in the last Ice Age. It is worth noting that, as in this case, isolated human bones have often been found around elaborate graves of the period, not only at Dolní Vestonice and nearby Pavlov but also at Predmostí and even farther afield at Sungir, near Moscow.

A rescue excavation in 1986, in a level at Dolní Vestonice whose loess deposits were being exploited by heavy machinery, led to the discovery of another camp of the Pavlovian period. A line of hearths extended toward the hills and contained large fragments of carbonized wood; there were also many tools of stone and bone. A radiocarbon date of 27,660 years ago was obtained from this level.

The upper part of a human skull (DV XI) belonging to a man about 40 years old was found here near a hearth, together with bones of mammoth, reindeer and wolf. A fragment of frontal bone (DV XII), probably from the same individual, lay 16 yards (15 meters) away and displayed a large wound above the right eye that had healed but that probably had lasting effects on the individual's health. It seems to have been caused by a hard, blunt instrument

BELOW The three skeletons unearthed in the same grave at Dolní Vestonice in 1986. The individual of ambiguous sex lies in the center with a definite male at either side.

wielded with some strength and may thus represent one of the rare clues about the occurrence of violence during the Upper Paleolithic.

On August 13, 1986, a remarkable triple burial was found only 16 yards (15 meters) from the skull and 33 yards (30 meters) from the frontal bone. The skeletons were in a shallow pit, in an extended position, side by side, with their heads toward the south. All three were young, between 17 and 20 years of age. The eldest, lying supine in the center (DV XV), was very gracile; this, together with the shape of the pelvis, points to a female. On her left was a big, robust skeleton, more than 5 feet, 9 inches (175 centimeters) tall (DV XIV, a 17-year old male), lying on its stomach, with its head turned away to the west; its left arm covered the female's hand, as if holding her. On her right was a smaller robust male (DV XIII), aged 18 or 19, lying on its left side and facing the female; both of its arms touched her pelvis. The three skulls were in a soil impregnated with red pigment (the bigger man's skull also had a coating of white powder), and the red crust around their frontals contained the remains of diadems of wolf and fox teeth and small ivory beads; fairly common in burials of this period, particularly in central and eastern Europe. There was also a considerable concentration of ocher under the female's pelvis and between her thighs. The pit contained a few small fragments of animal bone (reindeer and wolf), some flint flakes, an engraved pebble, and a cluster of unperforated shells.

A quantity of fragments of carbonized wood, up to 20 inches (50 centimeters) long and 4 inches (10 centimeters) wide, lay directly on the skeletons as well as near them. These may be the remains of some sort of tomb cover, a replacement for the covering of mammoth shoulderblades usually found on Upper Paleolithic graves in this area. It was perhaps burned as part of the funerary ritual, as the greatest mass of carbonized wood (a funeral pyre?) lay immediately west of the grave. The corpses themselves were not burned, but the presence of brick-red burnt loess immediately above the wood suggests that the fire was extinguished by having earth heaped over it.

## The Significance of the Group Burial

Wood from the grave has produced a radiocarbon date of 26,640 years ago. The three skeletons had clearly been buried at the same time (the female was put in first). Are they there because of an accident, an epidemic or for some other reason? One clue is that the female's bones show several pathological deformations. Her skull is strikingly asymmetrical and has a perforation from the frontal cavity to the left supraorbital torus; she had scoliosis of the vertebral column, deformation of the right iliac and an insufficient development of the entire right side and limbs. These conditions suggest that she may have contracted rickets or encephalitis at a very early age. In her mouth was a fragment of burnt reindeer pelvis,

**ABOVE** This thumb-sized carving from mammoth ivory, found at Dolní Vestonice and dating to c.24,400 years ago, is usually assumed to be a woman because of its elaborate coiffure. The asymmetry of its face has led some people to link it with the enigmatic individual in the center of the triple burial.

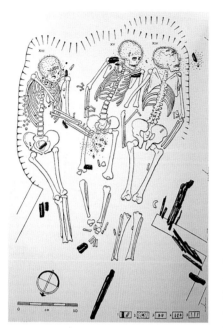

which may have been used as a clamp to bite on during times of great pain, as it shows traces of scratches and pressure.

One explanation, put forward by Klíma, is that the burial is a reproduction of a real event – a failed birth. The man on the female's left holds her arm to give comfort but cannot bear to look, while the man on her right – perhaps a medicine man – tries to help the delivery, but in vain. Klíma suggested that the ocher round the female's pelvis marks the original location of a fourth skull or the complete skeleton of a newborn child whose bones have disintegrated.

The idea is appealing, but the supposedly complete disappearance of the newborn baby's remains is made slightly doubtful by the good preservation and virtually complete state of the three skeletons. In addition, there are other enigmas – for example, the male with his hand on the female's pelvis was skewered to his sacrum by a large piece of wood, while the other male's skull was smashed (and further deformed by the weight of earth on it). It is possible that these two apparently healthy and powerful males met violent deaths, perhaps sacrificed to accompany the dead female – but some specialists remain unconvinced that the central figure is indeed a female rather than an effeminate, gracile male. This handicapped individual, like the Romito dwarf (see p. 148), adds proof to the already well-established notion that Ice-age communities took good care of, and perhaps even accorded special importance to, individuals with physical disadvantages.

ABOVE Drawing of the triple burial, showing (key: left to right) the charcoal fragments; the red color; a fossil shell; perforated teeth and ivory pendant; the grave's outline.

BELOW The importance of the central figure's pelvic area in the layout of these bodies suggests to some that the two men were killed owing to some kind of inadmissible behavior involving this crippled woman or effeminate male.

# Fenghuangshan Tomb 168,
## Jiangling, Hubei Province

The last three decades (from the 1970s to the present) can be seen as the golden age of Chinese archaeology. Most of the discoveries in that time have been salvage/rescue excavations that brought important and astonishing results to light, such as the jade shroud of Prince Liu Sheng in his tomb at Mancheng, Hebei province, dated to the Western Han Dynasty (206 BC – 6 AD), and the mausoleum of the First Emperor, with its terra-cotta army (third century BC).

ABOVE Despite having moved around in the tomb because of water penetration, many of the tomb's grave goods were found to be intact.

## The Tombs at Phoenix Hill

It is one of the great merits of the archaeological activities in the 1970s and 1980s that two unrobbed and fully intact Han Dynasty tombs were brought to light: Tomb 1 of Mawangdui, Hunan province, with the intact corpse of the noblewoman Lady Dai (see p. 51) and, only two years later, three tombs located at the so-called Phoenix Hill (*Fenghuangshan*) in the southeastern part of ancient Jinan city, the former capital of the Chu kingdom. Archaeologists first excavated the fully intact Phoenix Hill Tomb 168 as well as Tomb 167. Tomb 169 was partly destroyed. Tomb 167 consisted of a wooden chamber with three rooms, two for grave goods and one central room for the coffin. When the coffin was first opened, it was filled with a red liquid. After the coffin was emptied of this, layers of textiles and mats came to light. The textiles, such as silk and hemp, could be unwrapped layer by layer; finally, parts of a skull and hair were found. Scientific research proved that the tomb belonged to a woman, deduced from the exclusive grave goods to be a noblewoman, probably the wife of the owner of Tomb 168 – which was to prove a sensational discovery.

## Tomb 168

The tomb belonged to Sui Xiaoyuan, a high district official of the ninth rank (*wudaifu*) who held an office in Siyang, northwest of today's city of Jingzhou. According to inscriptions on the bamboo slips discovered, he was buried in the thirteenth year of the reign of Emperor Wen of the

**ABOVE** In the eastern compartment servant, horse and chariot models formed a funeral procession, which has been reconstructed and can now be seen in the local museum.

Western Han dynasty (167 BC). He found his last home only 30 feet (9.1 meters) away from Tomb 167 on the eastern slope of Mount Fenghuang.

The wooden tomb chamber was fully intact and placed in a rectangular shaft. It was divided into four rooms: two side rooms (*bianxiang*), one at the head of the coffin chamber (*touxiang*), and the main chamber with the double coffin. The outer and inner coffin were coated with black lacquer and tightly sealed with raw lacquer and linen. The inventory was recorded on bamboo slips, providing rich material for the nomenclature of the ancient grave goods. Wooden figures, such as servants and horsemen, and over 500 lacquered, ceramic and bamboo vessels were placed in the tomb, all originally wrapped carefully in silk or hemp cloth. The lacquers in particular were like new when they first emerged from the tomb; they constitute a body of material of great importance in the study of the esthetic aspects of Han artwork.

Most interesting was the discovery of an ink stone, a brush pen and some blank bamboo slips, among the earliest examples of writing materials. No ink had hitherto been found, but small fragments of ink pills were discovered in the tomb of Zhao Mo, who resided as King of Nan Yue in today's Guangzhou and died in 122 BC.

## The Corpse

In 1974, a comprehensive study of the corpse revealed that Sui Xiaoyuan was about 60 years old when he died. His corpse was 5 feet, 6 inches (168 centimeters) tall and weighed 116 pounds (52.5 kilograms). According to the Cinese archaeologists, the body was so instact that it looked as if the man was asleep, and the anthropological characteristics corresponded to the contemporary Han nationality from central and southern China. The skin was still soft, the soft tissues still elastic and the joints still flexible. All 32 teeth were intact but, astonishingly, no hair, fingernails or toenails were found. X-ray examination showed a complete skeleton. Despite shrinkage, all the inner organs were still in their place and clearly distinguishable, and some were perfectly preserved. Electron microscopic examination revealed that the collagen in the connecting tissue fibrils showed a regular microstructure hardly distinguishable from that found in fresh collagen fibers. Examination of the head showed that the brain was tumid and occupied more than three-quarters of the cranium. The myelin sheath of the nerve fibers maintained its typical lamellar structure, a condition unique in the study of ancient Chinese corpses. Investigation with light and electron microscopes showed that all the tissues, including the cartilage and the muscles, with their distinct light and dark bands, were well preserved. The cartilage cells and muscle fibers were also discernible.

ABOVE A model of a stove and cooking pots still contained firewood when unearthed from the side compartment.

A number of proteins were preserved in the muscles and brain tissues, including tropomyosin, one of the four important proteins in the muscle cells. An immunological study showed that the structure of the proteins was intact. The liquids in the brain, liver and skin contained a lot of cholesterol, and some of the tissues even showed traces of saccharides. The blood type was AB.

The conditions of preservation of this corpse are comparable to the female corpse in the Mawangdui tomb. Likewise, we know the diseases this man suffered from during his lifetime and the reason for his sudden death. He suffered from a chronic gastric ulcer that was complicated by an acute perforation of the stomach. He also had gallstones, which led to an acute cholecystitis (inflammation of the gallbladder). In the course of bacteriological research, two phyla of bacteria were detected: *Clonorchis sinensis*, also known as the oriental or Chinese liver fluke, and *Schistosomasis japonicum* – both small worms that infect the lining of the bile duct and cause an inflammation response in the surrounding liver tissue. Moreover, tapeworms and whipworms were found in the intestines. The examination of the artery walls and their lack of elasticity revealed a serious

BELOW Today the mummy of Sui Xiaoyuan is on display in the local museum in Jingzhou, Hubei province.

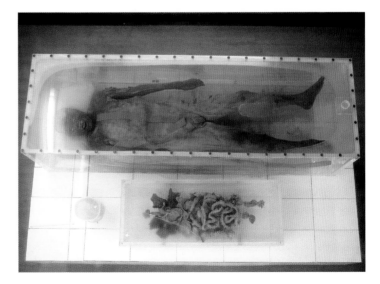

atherosclerosis. Further examination showed that a rupture of the stomach artery, among the above-mentioned problems, was responsible for extensive bleeding in the pleural and peritoneal cavities and inner organs. A breakdown of the whole system of the inner organs was the result, ending with a rupture and penetration of the diaphragm (diaphragmatic hernia).

Astonishingly, although nails, hair and fat deposits were lacking calcium, salt was widespread in the body, possibly as a result of the corpse's long period of immersion in the reddish-black coffin fluid.

## The Coffin Fluid and the Wooden Chamber

Sui Xiaoyuan's coffin was completely filled with a dark red acidic fluid. The corpse was placed in an extended position, facing east. At the bottom of the coffin archaeologists discovered textiles, cinnabar (*zhu sha*) ore and black beans. Analysis of the fluid showed that it functioned as a bactericide. The Chinese chemists set out to find whether or not the red liquid really preserved the corpse.

There are two other examples of tombs filled with liquid: the tomb of Lady Dai from Mawangdui in Hunan province, and Phoenix Hill Tomb 167,

**BELOW** Scientists of the Hubei Medical Academy carefully examined the corpse of Sui Xiaoyuan. The body had been astonishingly well preserved, and shed new light on burial practices in the Han Dynasty.

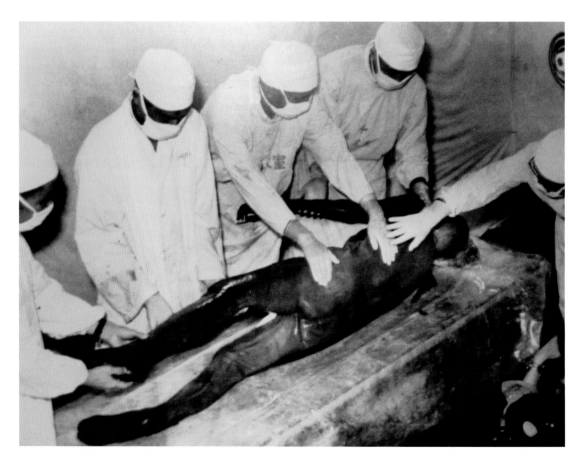

adjacent to Tomb 168. While the corpse of Lady Dai was extremely well preserved, in Tomb 167 only parts of the skull and chin remained intact. Assuming that a special recipe was known 2,000 years ago for producing a fluid that could mummify a corpse, why would it only sometimes have a preservative effect?

After some research, archaeologists concluded that a number of favorable circumstances finally led to the perfect preservation. First, the perfect geological, hydrological and climatic environment provided the context. Second, the tomb's layout provided optimal conditions for a long preservation. The tomb itself was dug as a rectangular shaft with vertical walls. Only a sloping pathway led to it, ending at the top of the wooden chamber. At the bottom of the shaft, sand and two wooden logs separated the large wooden chamber from the original soil. The wooden chamber itself was made of a hard and durable wood. The logs for the chamber were thick, and the joints were made only by mortise and tenon. A thick layer of white and gray clay was used as an airtight and waterproof seal, producing an average temperature of 61° to 68°F (16° to 20°C).

In the case of the Mawangdui tomb, an extra layer of charcoal was placed around the wooden chamber before the white clay was added. The effect of both methods seems to lead to the same result. Within the chamber, a central room held the double coffin. A thick layer of lacquer was painted on the inside and outside, while the edges and corners were reinforced by textiles glued with lacquer. The airtight seal provided the inside of the coffin with an aseptic environment, free of air and oxygen. Usually, these nested coffins fitted exactly into the coffin room, but in the case of Tomb 168, a lot of space surrounded the doubled coffin at both the sides and the ceiling. Despite the waterproofed seal, groundwater penetrated the wooden chamber as well as the various rooms. When the tomb was opened, the groundwater level was 29½ in (75 cm) high.

Research on the coffin itself revealed that, due to the water level in the room, the coffins could move and tip over. The coffin floated, and the thick lacquer skin did acquire some small ruptures. Water penetrated, and fluid also accumulated in the coffin. This could only have happened long after the entombment, and therefore the fluid could have played a role only in the later period of the long-term preservation.

But what about the cinnabar and the black beans? While the function of the black beans is not at all clear, cinnabar was widely used in Chinese history. It is a mercury salt that is acid- and alkali-resistant and cannot be dissolved in water. It can be mixed with hydrochloric acid and nitric acid. It was often added to raw lacquer, then used to coat drinking and eating vessels and coffins. In the *Shiji*, the Records of Sima Qian (145–85 BC), the great historian describes the tomb of the First August Emperor of China, Qin Shihuangdi (reigned 221–210 BC). Imitating heaven and earth, the tomb ceiling was painted with a starlit heaven, and a mercury river flowed through the tomb's architecture. Until now, no one has opened this tomb due to the costs involved, and only the huge terra-cotta army has

been excavated. In Daoist practice, cinnabar powder was used as an elixir for longevity. Today it is still used as an anti-irritation medicine and as a preventative against poisoning.

Burial practices are described in books about rites – for example, the *Liji* and the *Yili*. While the death ritual and the difficult rules for correct entombment are described in detail, the use of cinnabar as a method of mummifying the dead is not mentioned. Thus, it can be assumed that the red liquid is the result of an accidental mixture of ingredients.

## Tradition

For understanding conventions in mortuary customs, the most systematic data are provided by large burial sites dating to the earlier Zhanguo period around the old Chu capital of Jinan, close to today's Jiangling area. Among these, Mashan Tomb 1 (dated 340–278 BC), excavated in 1982, provided clues for understanding the tradition of the burial practice in Tomb 168, which was only a few kilometers away but more than 100 years later in date, as well as Mawangdui Tomb 1.

The construction of the tomb followed the design of the vertical pit tomb. A sloping passage led to the bottom, where the coffin chamber was placed. The wooden coffin itself was placed on two traverse beams, sometimes in grooves, at the base of the shaft, in order to keep the bottom of the tomb chamber away from the moist soil. The chamber was divided into three rooms and sealed with gray clay to prevent water penetration.

The caskets were lacquered, most frequently in black outside and red inside. A flat, rectangular, wooden board carved in openwork with geometric or zoomorphic patterns was placed in caskets under the deceased, and was probably used to hold the corpse while washing and preparing it for burial.

Openings of the corpse were sealed; for example, the mouth was filled with rice, as in Mashan Tomb 1, where a female of around 40 was buried. In this case, only the skeleton and some other parts remained; for instance, on her hands some dry muscle and tendon fibers still survived. Originally the hair was bound up under a cap and the feet placed in shoes. Fragrant bags were placed on the body. As with the Fenghuangshan and Mawangdui mummies, the Mashan woman's hands were tied together, then connected by a string to the feet. Then the corpse was wrapped in 13 layers of cloths, either of silk or hemp textile. In addition, the outer wrapping was tied with nine silk ribbons. The fully wrapped corpse was placed in the casket, which was then tightly bound with fiber strings. In the case of Mashan, as in Mawangdui, a fragment of a silk banner, very decayed, and a bamboo stick were found on top of the casket.

The burial techniques are the same in both the Mawangdui and Fenghuangshan tombs, and there was no groundwater penetration. Yet in Mashan tomb 1, the air- and watertight sealing in the coffin and the wrapping with silk cloths did not protect the corpse from decay.

# Turkana Boy:
## A 1.5-Million-year-old Skeleton

## Beating the Odds

Chances are stacked against the survival and recovery of the bones of early humans. For a start, they were rare creatures on the African landscape, and they did not bury their dead. Their corpses, even of those who did not succumb to predators, were quickly destroyed by scavengers and trampling animals, and the remaining bones crumbled through weathering and entanglement by vegetation. Occasionally, however, pieces of bone and, particularly, teeth survived long enough to be covered by sediments that protected them from the ravages of the open veld. Over time, minerals from the sediments seeped in and replaced their decaying organic materials until they turned to stone and became the fossil remains of once-living organisms. Then they wait – until their final resting place is exposed by erosion or excavation to the sharp eyes of a paleoanthropologist, a scientist who studies human evolution. The recovery of even a partial early human skeleton is rare; usually the remains are so fragmentary that simply trying to identify them can fuel lively debates among scientists..

## Hitting the Jackpot

However, luck was on the side of the paleoanthropologists who had pitched camp beside the sandy bed of the Nariokotome River some 3 miles (5 kilometers) west of Lake Turkana in northern Kenya one August day in

**RIGHT** Working under the hot African sun, the excavation team carefully sifts through the sediments at Nariokotome to recover almost all the bones of a 1.5-million-year-old early human: only his feet and a few other pieces were not found.

1984. While taking a stroll, the experienced eyes of Kamoya Kimeu spotted a scrap of early human skull bone lying among pebbles on the banks of a dry tributary at a site named Nariokotome III (NK3). The team, led by Richard Leakey, then of the National Museums of Kenya, and Alan Walker, an American colleague, had on countless previous occasions stumbled on fragments of early human bone that invariably turned out to be isolated pieces. This time, their initial complacency turned to disbelief and eventually ecstasy when the painstaking investigation of 1,962 cubic yards (1,254 cubic meters) of sediments at the site over several seasons led to the piecemeal recovery of virtually an entire early human skeleton.

## Recovering the Remains

Once the bones were exposed by careful excavation with dental tools and "Olduvai picks" – sharpened 6-inch (15.25-centimeter) nails attached to L-shaped pieces of wood – their position in relation to markers was noted. They were coated with plastic hardener and, in the case of the left thigh bone, encased in a plaster box, before being lifted. All the sediment was carefully washed through sieves to ensure that every scrap of bone was recovered. About 70 fragments of skull bones in addition to teeth, and 80 fragments of other bones belonging to a single early human skeleton, as well as numerous animal bones, were found and reassembled with glue, much as one would mend broken china. The skeleton was given the catalog number KNM-WT 15000 (KNM for Kenya National Museum and WT for West Turkana), and suffixes added as each piece was located and delivered to the museum in Nairobi. The skull is KNM-WT 15000 A, the lower jaw KNM-WT 15000 B, and so on to AA, AB, AC and, eventually, CB. The entire skeleton is also known as "the Turkana Boy."

## Identifying and Dating the Skeleton

KNM-WT 15000 compares favorably with bones of creatures with thick skulls, heavy brow ridges and massive, projecting faces that occupy a position on the human family tree between our first upright walking ancestors and ourselves. The first fossil of these to be discovered was a skull cap found by Eugene Dubois in Java in 1891. He gave it the scientific name *Pithecanthrppus erectus* – literally, "upright ape man," known today as *Homo erectus*. Depending on which features are thought significant for showing relationships between fossils on the human family tree, some researchers consider the Turkana Boy to be very like *Homo erectus*; others consider that he more closely resembles a similar African species, *Homo ergaster*, "the workman," named after a jaw bone found on the eastern side of Lake Turkana.

Luckily for paleoanthropologists, the Turkana Boy lived in a region of active volcanoes whose deposits can be dated. When ash or lava from a volcano has cooled, radioactive potassium 40 in the material begins to decay at a known rate to stable argon 40. By measuring the proportions of potassium 40 and argon 40 present today, scientists can determine how much time has elapsed since the rock formed, and, by inference, the age of fossils found in or between the rock layers. In recent decades, a refinement to the technique has involved measuring the ratio of artificially made argon 39 to argon 40. Both conventional potassium-argon and argon-argon dating were used to determine that the Turkana Boy was sandwiched between volcanic ashes respectively dated to about 1.88 million and about 1.39 million years ago. By using measurements of how distant the skeleton was from each of these layers and assumptions about how fast the intervening layers would have formed, an estimated age of about 1.53 million years was obtained.

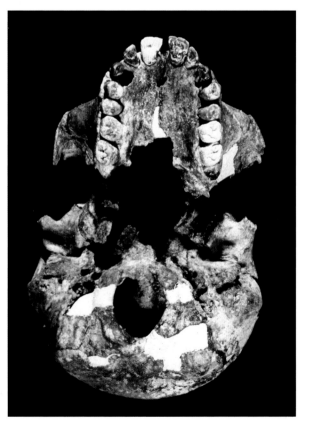

**BELOW** The base of the skull of the Turkana Boy. This was the first early human found with a skull sufficiently complete to provide an accurate measurement of the size of the brain.

## The Skeleton Speaks

### Age at Death

When the Turkana Boy died, his permanent teeth had not yet fully erupted; the lower jaw contains permanent incisors, canines and premolars as well as the first and second molars, but only some are completely formed, while the upper jaw still has milk canines. The third molars are absent. For the skeletons of living humans there are several ways of estimating

RIGHT The lower jaw of the Turkana Boy is very similar to remains of *Homo ergaster* from sites east of Lake Turkana. It provides the only known clue to the possible cause of his demise: a lesion on the right side indicates he suffered inflammatory gum disease.

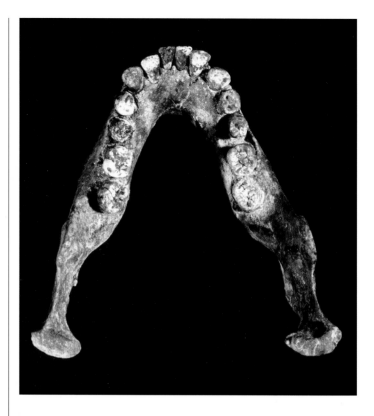

age using sequences and age ranges for tooth eruption and wear from many populations Had the Turkana Boy been a living human, some methods indicate he was 10½ to 11 years old, while other methods indicate an age of 11 to 12 years. On the other hand, ape scales of tooth eruption suggest a slightly younger age, about six to nine years, although the canines are too advanced in comparison with the other teeth by ape standards.

All long bones have an epiphysis, or separate center of bone formation, at one or both ends. The joining or fusion of the epiphyses with the midsection of long bones is another useful criterion for estimating the age of juvenile skeletons of living humans, because it occurs in known sequences and within certain age ranges. In the case of the Turkana Boy, most of the centers of bone formation had appeared but most of the epiphyses remained unfused, particularly those of the hip bones, although those of one end of the upper arm bone and part of the elbow joint were in the process of fusing. This state of skeletal maturation occurs between about 11 and 15 years in living humans but between only seven and eight years in apes.

A third feature that can be used to indicate the age of juvenile skeletons is height or stature, which is estimated from the length of the long bones. Using the Turkana Boy's thigh bone length of 17 inches (43.2 centimeters), he was about 5 feet, 3 inches (160 centimeters) tall at death, a stature typical of modern late adolescent 15-year-olds, and he would have been about 6 feet, 1 inch (185 centimeters) tall had he survived to adulthood.

The paleoanthropologists initially thought they had found an unusually tall individual, but reexamination of other early human remains from the same period confirmed that the Turkana Boy's kind were as tall as ourselves. Nevertheless, the skeleton probably represents the tallest early human known. Although it is difficult to estimate weight, by factoring in projected body breadths it seems likely the boy weighed some 106 pounds (48 kilograms) at death and could have reached about 150 pounds (68 kilograms) as an adult.

In terms of dental and skeletal development, the Turkana Boy is clearly adolescent but not a perfect match for either living humans or apes. Rather, he can be considered to have patterns of aging that fall between those of living apes and humans.

## Sex

The most reliable body part for identifying the sex of an adult skeleton is the hip girdle, because it has been adapted in females to facilitate childbearing – for example, by the presence of a wide, spreading greater sciatic notch, an identation found on the lower border of the hip bone. The narrowness of the Turkana Boy's greater sciatic notch clearly indicates that he was male rather than female. Maleness is also suggested by his tall stature and general robustness.

## Body Size and Shape

Modern humans who live in colder climates tend to have shorter limbs, while those living in more tropical regions have longer limbs, which result in more body surface area and hence promote heat loss. The tall, thin body and long limbs of the Turkana Boy are completely unlike those of earlier human ancestors but like those of modern tropical populations. They suggest he was adapted for living in a hot climate with extremely high mean annual temperatures of the order of 87.4°F (30.8°C). His body proportions are, in fact, almost identical to those of the tall Dinka people of southern Sudan, who live about 124 miles (200 kilometers) west of the Nariokotome River. However, the resemblance is physiological rather than genetic, as the Dinka are no more closely related to early humans than any other modern people.

Studies of the kinds of sediments, fossil pollen and animal bones indicate that the Nariokotome area has had the same hot climate as today for the past 1.5 million years. This means that if the Turkana Boy cooled off by sweating, as modern people do, he would have needed a great deal of water (the field crew who excavated him can confirm this!).

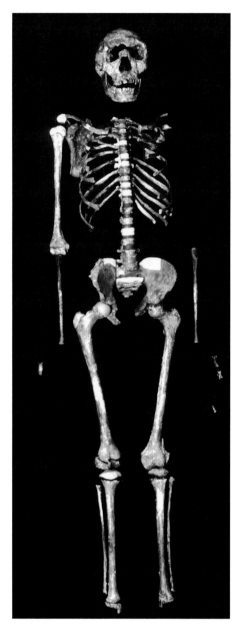

**ABOVE** It was once thought that early humans were small creatures and that humans grew gradually taller over the millennia. However, the long limbs of the Turkana Boy indicate that our ancestors had modern body proportions and were as tall as ourselves by 1.5 million years ago.

Sediments from the site indicate that shallow bodies of water appeared for short periods, such as one would find on a marshy floodplain that became seasonally inundated, 1.5 million years ago. The heat would also have caused the Turkana Boy to lose large amounts of salt, which he could have replaced by eating salt deposits or animal blood and tissues. The kinds of animal bones and how they were damaged at sites where early human stone tools are also present suggest that, although early humans consumed meat, much of it and particularly that from big game animals was probably more usually scavenged than hunted.

## Posture and Locomotion

The upright posture and bipedal, or two-legged, locomotion of living humans require the trunk to be balanced over the legs by a series of curvatures and other adaptations in the spine and hips. The curvatures of the spine, the orientation and balancing of the hips, and the presence of a barrel-shaped rib cage like that of modern humans rather than a funnel-shaped one like apes indicate that the Turkana Boy was fully adapted to habitual upright walking; his locomotion was strikingly like that of modern humans.

However, two notable differences exist between the spine of the Turkana Boy and that of modern humans: 96 percent of us have five bones in the lower back region of our spines, but the Turkana Boy had six. It is thought that the extra bone may have initially faclitated the development of the spinal curvatures necessary for effective bipedalism, but that it disappeared after the time of the Turka na Boy. His back bones also had relatively smaller spines than those of modern humans and enclosed a narrower canal for the spinal cord. This would have reduced the number of nerves to the chest, which may have in turn limited his ability to regulate air passing from the lungs to the mouth and hence prevented him from speaking as we do.

## Brain

Modern humans are distinguished by their relatively large, complex brains, which have an average volume of 82 cubic inches (1,350 cubic centimeters) in comparison with that of some 27.5 cubic inches (450 cubic centimeters) for apes. Despite the surprising similarities between the skeleton of the Turkana Boy and that of modern humans, his brain capacity of about 53.7 cubic inches (880 cubic centimeters) is only some two-thirds that of ours, which means that his behavior could have been very different from ours. The inside surfaces of his skull show the existence of areas of the brain associated with speech in modern humans, but these areas could have been involved with motor programming rather than language. Interestingly, the structure of his brain has a number of asymmetries typically associated with right-handed modern human males.

## How Did the Turkana Boy Die?

Although the floodplain grasslands around Nariokotome would have been home to a wide variety of animals, the completeness of the Turkana Boy skeleton strongly suggests that he was not the victim of a predator. The only hint of how he may have met his death is damage suggestive of a gum infection on the right side of his lower jaw. This could have happened not long before his death when he lost his second milk molar. Two small marks indicate that pieces of the roots were left behind and the upcoming permanent premolar had difficulty erupting. The resulting inflammation could have become infected, resulting in septicemia (blood poisoning), a common cause of death before the development of antibiotics.

The skeleton was found in a scatter of broken bird and mammal bones together with some remains of fish and aquatic reptiles. The teeth were clustered in a hollow about 9.8 feet (3 meters) away from the main concentration of bones. Almost all the bones are broken, but they are not weathered. This context suggests that when the boy died, his corpse either fell or was washed into a marsh, where it floated face down decomposing for a time. The teeth dropped out and were washed into a hollow probably formed by an animal footprint, while the body continued to drift slowly back and forth. It was trampled by large animals and, when only a few bones were still connected by ligaments, the remains were washed into a concentration in the shallows, where they became embedded in mud and remained until they began to erode out 1.5 million years later. The Turkana Boy's skeleton defied the odds and survived to provide us with a remarkably complete image of ancestors we previously knew only from isolated fragments of bone.

Despite his great antiquity his body, apart from his skull, is surprisingly like our own and shows that early humans had already reached our size 1.5 million years ago. A paleoanthropologist has even quipped that, if suitably dressed, the Turkana Boy could pass unnoticed in a commuter crowd, provided he concealed his low forehead and large brow ridges under a cap. However, his behavior could have made him stand out, and he would probably have had difficulty reaching his destination, as his kind had not yet evolved the large, complex brains that are our hallmark.

# Vilnius and the Ghosts of the
## Grande Armée

L ithuanian workers made a strange discovery in December 2001 on the hill of Siaures Miestelis, north of Vilnius, the capital. While laying pipes in the grounds of an ex-barracks of the Soviet army, they were surprised to come upon a multitude of skeletons, femurs, skulls, tibias, all piled higgledy-piggledy in a communal pit. The remains of the bodies of infantry and cavalrymen, plus shreds of uniforms of blue and green cloth, pieces of gaiters, buttons bearing regimental identification numbers and a 5-franc coin bearing the imperial profile of Napoleon, alerted the researchers who had been called to the site and revealed the importance of the discovery. The wretched remains of the dislocated corpses lying in the Vilnius pit were those of the French Emperor's soldiers – soldiers of the Grande Armée – perhaps even, among them, a few members of the Imperial Guard! Be that as it may, this is by far the biggest grave ever discovered of Napoleonic soldiers who died during the retreat from Russia in December 1812.

## The Russian Campaign

Begun in June 1812, the Russian campaign ended six months later in an appalling fiasco. The Army of the Twenty Nations comprised 614,000 men divided into 12 army corps led by marshals. The first lines, numbering 450,000 men, included Poles and their Legion of the Vistula – Bavarians, Austrians, Saxons, Westphalians, Croats, Swiss, Portuguese, Italians, Dalmatians, Dutchmen, Prussians and French, including 47,000 men from the Imperial Guard. The Army also moved with 90,000 horses and 25,000 wagons and carts. Even before battle was joined, almost a third of the men and horses had died due to the lack of water caused by drought. Some soldiers reported that they drank mud. Napoleon lost 10,000 men in battle at Smolensk, but he emerged the victor. However, he was left with only 130,000 men with whom to take Moscow, some 278 miles (448 kilometers) away. Before Moscow, they fought at Borodino, where the price of victory reached 30,000 dead. So it was an army of barely 100,000 men that entered Moscow on September 14, 1812.

## The Retreat from Russia

Napoleon spent five weeks in Moscow waiting for illusory negotiations with Tsar Alexander, his former ally in the 1807 Treaty of Tilsit. Despite the fall of Moscow, Napoleon was in no position to destroy the Russian forces – which, although smaller in number, were constantly giving him the slip. Threatened by the arrival of winter and that of two converging enemy armies – one coming from the north and the other not far from Turkey – Napoleon sounded the retreat on October 19, 1812. On a fine autumn morning, he left Moscow with fewer than 100,000 soldiers and some 40,000 stragglers – profiteers of all kinds who traveled in the wake of the Grande Armée.

By the time they reached Smolensk, the retreat had become a rout. Winter and its first snowflakes appeared three weeks early. Accompanied by this natural ally, the Russian army ceaselessly attacked the columns of survivors, who were numb with cold and hunger. Reduced to 50,000 poor wretches before crossing the Berezina, the survivors who later crossed the Niemen, the river that separated Europe from the Russian Empire, numbered scarcely 20,000 men.

## Crossing the Berezina

Few names in history are as evocative of total disaster as Berezina, nor synonymous with so many acts of bravery. The Russian army had seized the only bridge across the river, but the French discovered a ford some 11 miles (18 kilometers) upstream. Four hundred *pontoniers* (bridge builders) of the Corps of Engineers threw themselves into the icy water of the Berezina to erect two bridges on supports and enable thousands of their comrades to reach the other bank. Fifty thousand men died or were taken prisoner in the process, and fewer than 30,000 men survived to push on toward Vilnius. On December 5, Napoleon learned that a coup d'état had been attempted in Paris. He abandoned his troops and hurried straight back to France.

BELOW More than 3,000 infantrymen, cavalrymen, officers, hussars or members of the Old Guard had been lying for two centuries in the frozen earth of Vilnius, the Lithuanian capital.

## Excavating the Site

The archaeologists had to wait until the first fine days of March 2002, and more clement temperatures, before they could tackle the ground, which had been frozen until then. A French team known for its expertise in investigating communal pits – notably the grave pits of the great plague of Marseilles of 1720–1722 – went to the site at the request of Rimantas Jankauskas, the head of Vilnius University's Anatomy Laboratory. Despite a temperature of 21.2°F (–6°C), the team worked without a break to finish the excavation as quickly as possible and enable the city's construction site to resume work. According to Olivier Dutour, head of the biological anthropology service of the Faculty of Medicine at Marseilles, the bodies seemed to have been thrown into the pit in a manner reminiscent of the mass graves of Rwanda or Bosnia.

Using a planimetric excavation technique, the Franco-Lithuanian team managed to differentiate the skeletons. Hundreds of bodies were unearthed in rapid succession. From a total area of 766 square yards (640 square meters) and up to four layers of corpses, a stupefying density was recorded – 7 corpses per square yard ! More than 3,000 skeletons were extracted from their matrix of sand by the archaeologists.

**ABOVE**  After waiting several months for the ground to thaw, the archaeologists discovered soldiers' uniform buttons among the bones. Some still bore the regiment numbers of the Great Army.

**RIGHT**  A macabre harvest: the thousands of bodies exhumed by archaeologists in the pit in the *Siaures Mestielis* quarter were transported to Vilnius University's Anatomy Laboratory for study. This would be an opportunity to verify whether some of them carried typhus, as history claimed.

But who were these men? A few of the 40,000 dead that history has deduced for Vilnius? This question, among others, is what Dutour and his team of Marseilles paleopathologists are trying to answer. For them, quite apart from the historical aspect of the discovery, the sample provided by the Vilnius grave represents a unique opportunity to evaluate the state of health of young Europeans of the period: age, stature, oral and dental hygiene, food deficiencies, illnesses, epidemics. All this information will be deduced from the bones found in the Lithuanian pit. The study had just begun at the time of writing.

The Lithuanian authorities allowed the French team to work for a few weeks before the pit had to be filled in because of the construction going on. It is now up to the team to obtain the maximum information in order to analyze the bodies. Some cases are quite distressing. Some of the men were so exhausted that they died in a crouching position, frozen on their heels. One officer was still wearing his shako on his head, decorated with a red, white and blue rosette.

Historical chronicles tell us that in the final weeks of the retreat from Russia, the soldiers flooding back toward Vilnius, with Cossack bands at their heels, were the remains of a decimated army, exhausted by hunger, disease – perhaps even typhus – but especially by the cold; no winter had ever been as severe as that of 1812–1813. The temperature descended to

**ABOVE** Piled up higgledy-piggledy, the soldiers of the Great Army were collected together in a communal pit – one of the first pits ever discovered dating from the Napoleonic wars.

ABOVE Out of more than 400,000 men who took part in the Russian Campaign of 1812, barely 20,000 would return home. This disaster marked the start of the decline in Napoleon's reign.

–38.2°F (–39°C). In December 1812, those of Napoleon's army who managed to reach the suburbs of Vilnius (known as the "Jerusalem of the North" and the "City of a Hundred Synagogues" because of the high Jewish population at the time), were half-starved, worn-out ghosts. Above all, they were survivors of the crossing of the Berezina, achieved under the most terrible conditions.

Though all of them were starving, many were also wounded or dying. During the Napoleonic wars, the most appalling fate was to die in a hospital. In that period, it was far better to die on the battlefield than to suffer agonies in the clinics, which had absolutely no resources – no anesthesia for amputations, no medicines for infections or gangrene. Most of the soldiers who died during the Empire's wars did so in a hospital. At no point was there time to take care of the wounded. The most able-bodied had to advance as best they could; the rest found themselves in a hospital – and died there. This is what happened at Vilnius.

By February 1813, two months after their arrival and that, almost immediately afterward, of their Russian pursuers, those soldiers who could not continue their march were taken prisoner. When the Russians entered the city, they feared an epidemic and tried to get rid of the corpses as rapidly as possible. Because the inhabitants protested against cremation, which caused a stink in the city, it was decided to bury the bodies in trenches. So the dead and dying from the hospital of Vilnius were thrown out of the windows. Even the body of General Lenormand, who had died in the hospital, was thrown into the courtyard along with the other corpses.

Of the 1,400 sick men who were in the hospital of Werka, near Vilna, on January 22, 1813, only 128 were still alive in March that same year. Of the 120 sick prisoners in the hospital of the Sisters of the Infant Jesus when the Russian army entered it on December 8, 1812, only 27 were still alive four months later. Such is the sad report of the few survivors.

Today, those in charge of the grave project are wondering what to do. The bodies are those of French soldiers, of course, but also soldiers from the Army of the Twenty Nations. Should the bones be reburied in a Lithuanian cemetery at Vilnius (the modern town of Vilna), or should a commemorative monument be erected? No decision has yet been made.

Few monuments commemorate the victims of the Napoleonic wars. A few stelae are in Egypt, a few on the Battlefield of Borodino, another is at Brilli, on the right bank of the Berezina, another at Borisov, and one more at Tilsit. A few generals' tombs exist, too, but hundreds of thousands of families were never able to mourn properly because their loved ones were simply termed "missing" or "prisoner" in the army's records. Perhaps the Vilnius site presents the opportunity to recognize those tens of thousands of forgotten soldiers who perished during that fatal winter.

**BELOW** Emaciated, exhausted, haggard, blackened with frostbite – an army of ghosts reached the suburbs of Vilnius in December 1812. For their part, the Russians also lost more than 300,000 men.

# Kennewick Man

W hen Will Thomas and Dave Deacy showed the human skull they'd found on July 29, 1996, in the American town of Kennewick, Washington, to police, they had no way of knowing the controversy it would cause. The resulting legal and intellectual battles would involve concepts of race, ideas about human colonization of the Western Hemisphere, the role of science in American society and even the application of federal law.

Police turned the skull over to the county coroner. More bones from the skeleton were soon found in the same area, eroding out of the bank of the Columbia River. Forensic analysis and identification of the bones were entrusted to James Chatters, a local archaeologist who usually did this work under contract for the county coroner. Preliminary analysis, based on Caucasoid characteristics of the skull, suggested it was not American Indian but more likely European. A second expert opinion confirmed this conclusion, suggesting the remains were those of an early settler.

The surprise came when the bones were dated. A stone spear point was found embedded in the hip of the skeleton, but it provided little help in dating, as spear points of this style had been used in the region over a 9,000-year period into historic times. Radiocarbon (carbon 14) samples successfully dated the bones to between 9,330 and 9,580 calendar years, or 7330 to 7580 BC, indicating that the bones were 9,500 years old. This proved Kennewick Man to be of Paleo-Indian age and one of the oldest prehistoric individuals to be found in North America. However, the bones showed little resemblance to American Indians.

Before additional studies could be performed, the U.S. Army Corps of Engineers (as the responsible federal agency, because the bones had been found in a riverbank), acting in concert with at least one American Indian tribal group, stepped in to stop the analysis and confiscate the remains. The Corps of Engineers also indicated that it intended to repatriate the remains to one of the groups claiming them for reburial, despite the lack of required evidence of cultural affiliation.

Under a 1990 federal law known as the Native American Graves Protection and Repatriation Act (NAGPRA), human remains determined to be American Indian are to be returned, after they have been studied, to their descendants based on tribal affiliation. This law applies to American Indian remains only. Any human remains found in North America that predate European contact (AD 1492), of whatever age, are also considered American Indian by process of elimination.

Application of this law, even within federal agencies, has been uneven across the United States. In many places, discovered human remains are successfully studied and tribal affiliation for repatriation is determined based on artifact identification, historic documentation, oral tradition and discussions with area tribes. In other cases, however, human remains of any age have been returned to any American Indian group claiming them regardless of tribal affiliation, based on a concept of Pan-Indian culture. Tribal groups have used NAGPRA to fight the federal government, pressing land claims and overturning treaties. Frequently, American Indian religious beliefs are used as arguments to deny the validity of Western science in much the same way the beliefs of creationists are.

A number of forces, including the state of Washington's congressional delegation, the world press and the scientific community, joined in an effort to delay reburial of the Kennewick Man remains until additional studies could be completed. When this approach failed, a group of prominent archaeologists filed suit against the Corps of Engineers, alleging that their decision to repatriate Kennewick Man was arbitrary and capricious and, among other things, denied important scientific study that could be of major benefit to the United States. In a successful move to have this lawsuit put on hold, the Corps of Engineers agreed to have a similar range of analyses conducted by its own scientists. Once these studies were completed, the Corps of Engineers buried the site during a congressional recess despite orders from the U.S. Congress not to alter the existing riverbank.

The U.S. Department of the Interior, in a political decision, declared Kennewick Man affiliated with all five of the American Indian tribal groups who originally claimed him and announced he would be repatriated to them for reburial. The government explained that NAGPRA

RIGHT Following the discovery of Kennewick Man's skull, additional portions of his skeleton were recovered where the bones (including this one) had eroded from the riverbank.

was an Indian law and that, in any doubt, decisions should go in the Indians' favor. The lawsuit by archaeologists, which had been on hold, was reopened. This legal battle continues today.

## Who was the Kennewick Man?

Completed analysis of the bones shows Kennewick Man to be a male in his forties. He was approximately 5 feet, 8 inches (173 centimeters) tall and probably weighed about 159 pounds (72.2 kilograms). He was well nourished on a diet primarily comprising fish, according to the carbon isotope composition of his bones. However, life had been tough for

Kennewick Man. During his adult life, he had received several blows to his head. A blow to his ribs had given him a flail, or collapsed, chest. He had never recovered from early injuries to his left shoulder and elbow. A large stone spearhead from an old wound, embedded in his hip, probably gave him constant pain.

Dating by the government's team confirmed the date obtained earlier. Geological studies conducted at the site of Kennewick Man's discovery provided additional confirmation of age. Attempts at DNA identification were unsuccessful, and earlier DNA studies had been stopped by the Corps of Engineers.

Kennewick Man does not resemble American Indians. Major differences are observed when his profile outline is placed over the skull of an American Indian. Kennewick Man has a narrow, lightly constructed, square jaw with a more prominent chin, a high-bridged projecting nose and a forward facial placement. An illustrated reconstruction was done of Kennewick Man. The resulting face has been interpreted as resembling English actor Patrick Stewart. Others have suggested a resemblance to the historic painting of Chief Black Hawk (AD 1767–1838, leader of the Sauk and Fox tribe in the western Great Lakes region) and his son.

It has been argued that racial terms such as "Caucasoid," "Mongoloid" and "Negroid" are superficial groupings that fail to adequately represent the variability that exists between individuals within each group. Using this argument, it has been claimed that Kennewick Man represents an extreme example of a type of individual present within any American Indian population. However, based on the statistical probability of such remains being found, it is more likely that he is a representative example of his population group. But what group would that be?

Multivariate statistical analysis of Kennewick Man, based on a series of skull measurements, showed him to be atypical of any modern people, but closest to Polynesians in some respects and the Ainu of Japan in others. Resemblances to American Indians were minor. Repeated sets of analyses found a similar pattern: Kennewick Man was closest to Polynesians and the Ainu, but unlike American Indians or Europeans. Analysis of dental characteristics yielded analogous results.

Kennewick Man, through his skull and dental characteristics, does not exactly resemble any modern population, including American Indians. Could he have been part of a discrete population that later became

**ABOVE** Controlled examination of the riverbank resulted in the recovery of even more of Kennewick Man's bones. The exposed stratigraphy of deposited layers of dirt and sand also helped date the remains.

**LEFT** This large stone spearhead was found embedded in Kennewick Man's hip. Notice the curved area, centre-right, where the bone has grown around the spearhead.

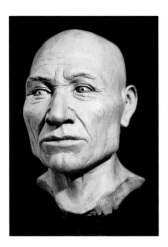

ABOVE This reconstruction of Kennewick Man exhibits the high, forward-projecting face, and square jaw with prominent chin. These are all traits that are not characteristic of American Indian populations.

extinct, either through death or intermarriage? A number of cultural traditions seem to have coexisted in North America prior to 9000 BC, when they begin displacing each other on a regional scale. Was this change in cultural traditions accompanied by a replacement of the population? This question has not usually been addressed. In North American archaeology, changes in cultural traditions have customarily been interpreted as representing cultural progress within a single population. Differences in populations have been considered minimal and single regional cultures the norm.

Unlike in other areas of the world, there has been little interest in tracking movements of discrete populations within the Western Hemisphere. Although early humans are assumed to have arrived in a series of waves, scientists have simply postulated continuity between the first ancient humans who crossed into the Western Hemisphere and modern American Indians. It has been assumed that a small genetic base led to a broadly homogeneous native population. Kennewick Man is a tantalizing indication that this was not the case.

The prehistoric human peopling of the Western Hemisphere may have consisted of multiple waves of small distinct groups entering the continent, possibly in a variety of ways. That these groups ultimately merged in isolation from the rest of the world is not in doubt. Nor is the fact, demonstrated by Kennewick Man, that populating the continent was more complex and involved than anyone has suspected. Some populations obviously prospered, while others seem to have died out. The few Paleo-Indian remains that have been found are rare discoveries containing valuable information about these earliest inhabitants. But will studies of these remains be carried out in the future? With Kennewick Man still in the courts, we await the final outcome.

RIGHT Portrait of Chief Black Hawk (of the Midwestern Sauk and Fox tribe), and his son. His narrow facial features (unusual among American Indians), are said to resemble those of Kennewick Man.

# CHAPTER THREE

# deliberate deaths

I n some cases skeletons themselves reveal scarred bones and wounds from missiles that betray death from foul play. Intact bodies also bear graphic witness to the various grisly ways in which prehistoric people have met with an untimely fate.

# The Iceman Reveals
## Stone Age Secrets

On Thursday, September 19, 1991, a pair of German hikers in the Alps made the most remarkable archaeological discovery of the late twentieth century. Protruding from a slush-filled gully along the border between Italy and Austria was the head and torso of a male corpse. The Iceman, as this individual has become known, continues to yield new evidence over a decade after his discovery.

### Recovering the Body

An Alpine rescue team arrived to recover the corpse, assumed to be a hiker lost in the previous few decades. They used a pneumatic hammer to free the body from the ice, damaging its left hip. Luckily, they soon ran out of electricity. Over the next couple of days, visitors to the site tugged and poked at the corpse, damaging it still further. Finally, it was freed from the ice, using ice picks and ski poles, and transported to Innsbruck in Austria.

Some of those involved with the body's recovery sensed he might be older than originally thought. A metal ax in a wooden handle found near the corpse provided a key clue, as did other artifacts of wood, leather, flint and grass. Over the next several days, the Iceman's age was extended backward by speculation, first to medieval times, then to the Iron Age. Finally, Konrad Spindler, professor of archaeology at the University of Innsbruck, inspected the corpse and its associated finds. His initial impression was that the Iceman was about 4,000 years old, dating from the Bronze Age.

Spindler's assessment caused an immediate sensation, for it now appeared that the Iceman was one of the oldest bodies ever recovered with so much flesh preserved. The Iron Age bog bodies of northern Europe (see p. 98) had demonstrated the scientific potential of prehistoric corpses, but the Iceman was far older. A preserved body several thousand years old had enormous potential for scientific investigation. Spindler and his colleagues set in motion one of the most dramatic detective stories in the history of archaeology.

**BELOW** The Iceman was found on a high Alpine ridge on the border between Austria and Italy. When the late summer sun thawed the ice around his body, it was spotted by Erika and Helmut Simon.

## Archaeology, Dating and Preservation

Researchers returned to the slushy gully and found more fur and leather, clumps of matted grass, a birchbark container and a net made of grass. None of these items could have found its way to the spot naturally. A bow had been found originally with the corpse; now a quiver containing arrows, some with feathers still attached, was recovered. The onset of winter soon ended these initial investigations.

Two important things had to be done immediately with the body. First, it had to be preserved in a way that did not destroy important biochemical evidence. Second, it had to be subjected to more precise dating, using the high-precision form of carbon-14 dating with an accelerator-mass spectrometer. The Iceman was placed in a sterile refrigeration chamber at the University of Innsbruck and swabbed with carbolic acid. Tissue and bone samples were sent to radiocarbon laboratories in Oxford and Zürich, while additional samples from the matted grass were sent to labs in Uppsala and Paris. Meanwhile, the delicate artifacts were sent to Mainz, Germany, where a team of archaeologists and conservators set about cataloging, preserving and restoring the finds.

The first carbon-14 dates caused a further sensation. All the dates from the different laboratories pointed toward an age of about 5,300 years, over a millennium older than the 4,000 years estimated by Spindler. But what about the metal ax, originally thought to be made of bronze, an alloy of tin and copper? Metallurgical testing indicated that instead of bronze, the ax was made from almost pure copper, consistent with the age indicated by carbon-14 dating.

## Copper Age Europe

The Iceman was not a Bronze Age person at all. Instead, he belonged to an earlier period known to archaeologists as the Neolithic or New Stone Age. During the late part of the Neolithic period, copper came into use, especially in southern Europe. Some archaeologists refer to this time as the Copper Age, although people continued to rely on stone tools as they had before.

During the Copper Age, or Late Neolithic, many types of pottery and stone tools were made. Burials often included lavish offerings such as large copper axes and ornaments of copper and gold. People lived in hamlets composed of self-sufficient households. Some households acquired greater status and wealth than others, and thus archaeologists can discern the roots of social differences during this period. These differences in wealth may have been due to access to resources such as copper as well as to the increased use of livestock for meat, wool, milk and power. The first evidence for wheeled vehicles and plowing appears in Europe in this period.

The Iceman lived and died during a fascinating time about which archaeologists wished to learn more. In fact, the items that he was

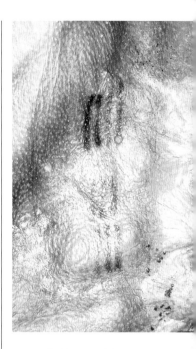

**ABOVE** Ötzi has over 50 tattoos, with the larger ones on his back, ankles, knees and calves. Although it is more likely that they were decorative, some have suggested that they had a therapeutic effect on his aches.

carrying and wearing made it clear how little they really knew about Copper Age life. Nothing had been known before about clothing and equipment made from perishable materials. The most fascinating questions involved the Iceman himself: Where did he come from? What was he doing in the Alps? Where was he going? How did he die?

## The Iceman Becomes a Celebrity

The establishment of the Iceman's age, the good preservation of his corpse and the unusual finds that accompanied him made him a natural subject for media coverage. Each new revelation seized the imagination of the public, and he began to take on a personality. New names were offered for him. "Similaun Man," as he was found hear an Alpine peak called Similaun, did not gain wide acceptance. Neither did "*Homo tirolensis*," or Tyrolean Man. The most popular nickname for the Iceman is "Ötzi," which refers to the Ötztal, a valley leading north from the site. For some reason, Ötzi stuck and appears in many popular publications.

Fascinating details about the Iceman's corpse began to emerge. When alive, he was about 5 feet, 3 inches (160 centimeters) tall and weighed about 110 pounds (50 kilograms). He was relatively old for a Copper Age person, probably between 40 and 53 years of age. Initial X-rays revealed that he was missing a twelfth pair of ribs, a rare but harmless anomaly. Of particular interest were tattoos on his back, knees, ankles and left wrist. Charcoal had been rubbed into small cuts, which then healed with the charcoal fragments still embedded in the skin. The tattoos occur as groups of lines a few centimeters long and, on one knee, a cross.

Initially, the Iceman was presumed to have been found on the Austrian side of the frontier. The border is not clearly marked, however. Media coverage of the find led to closer scrutiny of its location, and a survey established that the location of the find was 101 yards (92 meters) over the frontier on the Italian side. An arrangement allowed the Iceman to remain in Innsbruck for further study; he was brought back to Italy once a facility was built to receive him.

**RIGHT** Among the most unusual finds with the Iceman were two pieces of mushroom, *Piptoporus betulinus*, that had been perforated and threaded on leather thongs. Some have speculated that this fungus may have had some medicinal use.

## The Investigation Continues

In 1992, archaeologists returned to the site where the Iceman was found and examined the gully thoroughly. They recovered additional pieces of fur and leather as well as pieces of Ötzi's own skin, muscle, hair and even a fingernail. One remarkable additional find was the fur cap that the Iceman had been wearing.

Meanwhile, in Innsbruck, the Iceman's corpse was studied further. The fingernail showed that his immune system had suffered nutritional stress before his death. When he was alive, his nose and a rib had been broken and healed. Other ribs had been broken around the time he died. The Iceman's teeth were heavily worn, probably from eating grain milled on grinding stones, although he had no cavities. He suffered from arteriosclerosis and osteoarthritis. His toes had been repeatedly frostbitten. Smoke from campfires had blackened his lungs, and traces of minerals from metal smelting were found in his hair. By any measure, Ötzi had lived a hard life.

In Ötzi's intestine was a mixture of wheat and meat, along with other species of plants. This provided clear evidence that the Iceman came from an agricultural community that grew wheat, although it also showed that these farmers included other plants in their diet as well. Thirty types of pollen grains were found in Ötzi's intestine. The predominant species were hop hornbeam, hazel, birch and pine. Hop hornbeam requires warmth, so its presence is clear evidence that the Iceman had been in one of the warm valleys south of the Alps not long before he died.

Moreover, the hop hornbeam pollen provides a clear indication of the season in which Ötzi died. Early in the investigation, the prevailing assumption had been that the Iceman had perished in an autumn or winter blizzard. Hop hornbeam, however, flowers in the spring, as do the other species whose pollen was identified. The excellent preservation of the pollen grains indicated that they had been ingested not long before Ötzi had died. The Iceman had met his death at the end of winter, not at the beginning.

## The Iceman's Clothing

The items found with Ötzi provide an unprecedented glimpse into prehistoric life 5,300 years ago. The finds can be divided into two categories: clothing and equipment. Prehistoric clothing is rarely preserved, and none had been known from the European Copper Age. Similarly, Ötzi's equipment, much of it made from

**BELOW** Ötzi's clothing and equipment have been reconstructed for this display in the South Tyrol Museum of Archaeology in Bolzano, Italy. Clearly visible are his grass cloak, bearskin cap and insulated shoes. The actual artifacts are displayed in nearby cases.

perishable materials, had also been previously unknown to archaeologists.

Ötzi's clothing consists of his cap, coat, leggings, belt with a small pouch, loincloth, shoes and cloak. Perhaps the most unusual item is the sleeveless cloak made from bundles of grass stalks bound together with grass twine. Its original length was about 35 inches (90 centimeters), and it would have been worn over all the other garments. Such a cloak was probably warm and water-repellent. In fact, Alpine shepherds in recent centuries are known to have worn similar grass cloaks.

The coat, leggings, belt, loincloth and parts of the shoes were made from leather. It is remarkable how many species of animal contributed their skins to the Iceman's attire. His coat, leggings and loincloth were made from goat-hide, while the cap and the soles of his shoes were made from bearskin. Deer-hide was used for the upper parts of his shoes, while the belt and pouch were made of calfskin. It is surprising that he was not wearing anything made from wool, although fragments of textiles are found at other European sites from this period.

Each of Ötzi's articles of leather clothing has some unusual feature. His coat is a patchwork of smaller pieces of goatskin sewn together with animal sinews. The leggings were leather tubes that covered the thighs and lower legs; a deerskin strap was attached to the shoes to keep them from riding up. Ötzi's belt was a piece of calfskin measuring 6½ feet (2 meters) long that probably went around him at least twice, with another piece of leather sewn on to make a pouch. The loincloth was an apronlike garment held up with the belt. Two leather straps tied under the chin held the bearskin cap on Ötzi's head.

The Iceman's complicated shoes consisted of an oval leather sole, a net of twisted grass that encircled the foot, and a mass of grass insulation held in place by the net. The bearskin sole was held in place with a leather strap, with the net between it and the foot. Grass or hay was stuffed into the net, with a piece of deerskin across the top. Putting on these shoes required effort, so the warmth they provided must have been worth it, although their durability and traction probably left something to be desired.

## The Iceman's Possessions

The Iceman probably knew exactly what he might need for a trip across the Alps, and his equipment was carefully chosen; many different types of wood were found among his belongings. He also carried tools to repair his possessions. There are, however, puzzling aspects to the finds, particularly the unfinished nature of his bow and arrows.

The Iceman's ax was highly informative. Not only did its composition of nearly pure copper confirm the dating of the

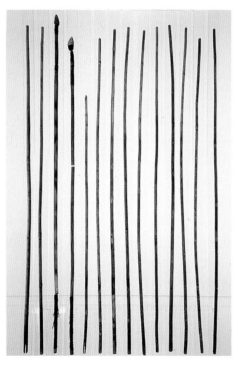

**BELOW** Ötzi's 14 arrows – two finished, 12 unfinished – were made from viburnum, a very resilient and straight wood. The shafts were notched at one end to hold stone or antler arrowheads, while at the other end they would have been fletched with feathers.

find but also its wooden handle shows how the metal ax was mounted and held. The handle is made from a yew branch close to 2 feet (60 centimeters) long, with a shorter branch extending from one end. The short branch was split to receive the copper ax blade, which was glued in place with birch pitch. The forked branch holding the blade was then wrapped tightly with leather straps. This multipurpose tool, useful both for cutting wood and as a weapon, was probably a prized possession of the Iceman.

Ötzi carried a backpack made from wood and leather. A hazel stick about 2.2 yards (2 meters) long, bent into a U-shape, formed a frame that was covered by a leather rucksack. He was also carrying a quiver made from chamois-hide that contained 14 arrows made from viburnum, a very tough wood. Only two had points, although the quiver also contained some tips of antler that possibly were intended for use as points. The Iceman's yew bow was also unfinished, leading to speculation as to whether he was still working on it when he died.

In his pouch and backpack, Ötzi carried many useful items. Two birchbark containers appear to have held embers for starting a fire. A net made from grass twine would have been useful for catching birds or small animals. A dagger with a flint blade and an ash handle was in a sheath made from plant fibers. The flint has been traced to a deposit in northern Italy, near Lake Garda. A rod of linden, pointed at one end with a fire-hardened piece of antler embedded in the point, was probably used for the detailed working of flint blades.

Some of the finds are mysterious. Two round pieces of birch fungus were attached to leather strips and may have served some medicinal purpose, perhaps as an antibiotic. A marble disk with a hole in the center was attached to a strip of leather, which in turn was strung onto a tassel of leather thongs.

## How Did Ötzi Die?

A recurring question about the Iceman is the manner in which he died. Initially, he seemed simply to have been a traveler caught in a snowstorm who died of exposure. Even during the spring, a sudden storm is possible, and when trudging through the remaining winter snow at this altitude, a traveler could easily become bogged down and die from hypothermia.

Ötzi's broken ribs and broken equipment, however, inspired further speculation. Konrad Spindler hypothesized that Ötzi had been fleeing from an altercation that had resulted in breakage to his arrows and his

**ABOVE** The Iceman's copper ax was set in a handle made from the trunk and branch of a yew. Some believe it was a ceremonial mark of authority, because the delicate handle might have broken from heavy use.

**BELOW** The small arrowhead found through CT scans in Ötzi's shoulder is thought to have entered from behind at the location of this wound, leading to speculation that he had been murdered.

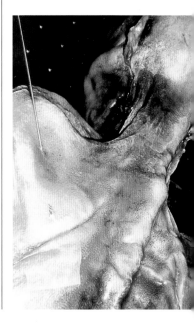

ribs. Alone and hungry, the wounded Iceman had sat down in the gully to die. Yet a bad fall could also have accounted for the breakage.

Recently, CAT scans revealed a small flint arrowhead embedded in the Iceman's left shoulder. Was this the cause of his death? Although there appears to be a mark on Ötzi's back that may be a trace of an entry wound, it is odd that no part of an arrowshaft was preserved. We know Ötzi lived a hard life, and that conflict was probably part of it, but it is still too early to conclude that he was murdered. The Andean explorer Johan Reinhard proposes that Ötzi was a sacrificial victim, like the mummies of Inca children found in the Andes. Reinhard offers no concrete evidence for this, instead speculating that the Iceman was a person of high status and that the nearby peaks were sacred to Neolithic people.

The most recent discovery has been of two deep wounds on Ötzi's right hand and wrist. It appears that a sharp object penetrated the base of his right thumb, causing serious injury, not long before Ötzi died. Yet if he was defending himself from a knife-wielding attacker, as some have suggested, why are there not more slashes on his forearm from parrying the blade, or puncture wounds to vital organs?

Melodramatic speculation that the Iceman was a fugitive, a murder victim or a sacrifice is premature. At the moment, it seems most likely that Ötzi died alone, a victim only of the terrain and the weather. His survival was impeded by his advanced age and the cumulative effects of his hard life. In the end, the Iceman's importance to archaeologists is not how he died but rather the evidence that his body and his equipment provide about how he lived.

**BELOW** Not far from the display of the artifacts he carried, Ötzi lies in his climate-controlled room. His empty eye sockets stare blankly upward as the visitors file past, paying their respects as if at a wake.

## Visiting Ötzi

On January 16, 1998, Ötzi and his possessions were transported to the new South Tyrol Museum of Archaeology in Bolzano, Italy. Ötzi's corpse is kept in a chamber with constant temperature and humidity. Visitors file past in silence to view his body through a small window. In the surrounding rooms, the numerous finds are displayed, while a full-size reconstruction of Ötzi in all his regalia gives an idea of what he looked like while alive. Outside the museum, Ötzi souvenirs and even "Ötzi pizza" can be purchased. One can only marvel at how an unknown traveler who died 5,300 years ago has become not only a major scientific revelation but also a world-famous celebrity.

# The Butchered Anasazi

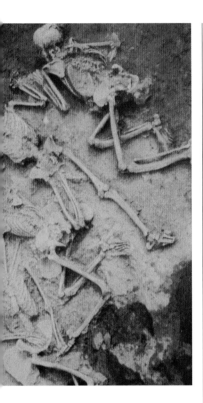

**ABOVE** An Anasazi mass grave from the Four Corners area. Shortly after death, these Anasazi were thrown into a subterranean kiva. All skulls were partially crushed and, on some individuals, limbs were removed.

he Anasazi were ancestors of the Pueblo Indians and lived in the Four Corners region of the American Southwest from about AD 600 until European contact. Mass graves are a common feature of Anasazi sites; they appear as collections of human bones, usually found within a contained area. They can occur in surface rooms, pit structures and kivas (subsurface ceremonial rooms or structures). They are also found in storage pits, rock shelters and naturally occurring rock cavities and crevasses. In rare cases, they have been found in the open.

Until recently, these graves were not even considered burials. When piles of broken and battered human bones were found, they were often dismissed as "disturbed deposits." Only formal burials, at times containing grave goods, were believed to represent Anasazi interment. Who were in these mass graves, and what did the graves represent?

We now know these are expedient mass graves containing the victims of tribal violence, reflecting the aftermath of tribal warfare. Within the mass graves, human bones are present as piles of disarticulated and partially articulated skeletons of numerous individuals. Skulls are often crushed and smashed into small fragments. Long bones are usually shattered and split, and ribs and other bones are broken into small pieces. Cut marks are present on some bone fragments (usually vertebrae and long bones), and some of the bones are burned. Some portions of the skeletons are usually missing.

## Modern Interpretations

It is now obvious that a number of long-held theories about Anasazi culture were wrong. A broader, more realistic view of Anasazi culture has developed recently as archaeologists have come to recognize and accept the presence of both violence and warfare within this culture. An expanded view of burial practices has also been accepted as we realize that the dead were dealt with in a number of ways. Formal burial was just one of several types of burials, although the total range of Anasazi burial practices has yet to be completely understood.

There has been some hesitation in accepting these new views of Anasazi society. It has been erroneously claimed that the mass graves are the remains of cannibalistic feasts. Raiding parties from Mexico were said to have traveled through what is now the southwestern United States, killing and then consuming their victims. This narrow view of Anasazi culture was a byproduct of the lingering belief among archaeologists working in the region that the Anasazi were peaceful egalitarian farmers.

**ABOVE** The threat of war forced many Anasazi to live in large, easily defended, communal structures, known as cliff dwellings. These were built in huge rock overhangs located midway up high cliff walls.

Warfare was considered rare among these people and violence a form of "abnormal or deviant" behavior. Using taphonomic models of bone breakage patterns developed for faunal analysis (taphonomy is the study of the processes that affect the completeness and composition of an animal's skeleton after death), proponents of the cannibalism theory approached analysis of these human remains as if they were butchered animals. Apparently similar patterns in bone breakage and in the occurrence of cut marks on the remains of butchered game animals and of warfare victims are claimed as proof that both activities are the outcome of the same motivation: human consumption. This approach is logical only if the premise is accepted that humans are simply animals. When cut marks appear on animal bones, they are the result of cutting and butchering for consumption. Therefore, this argument goes, cut marks on human bones must also be the result of consumption. In one study, supposed human fecal material containing human tissue was reported as definitive proof of Anasazi cannibalism and as supporting evidence for earlier such claims.

To date, however, not one of the claims of Anasazi cannibalism has been substantiated. Most champions of Anasazi cannibalism base their arguments on unrealistically narrow concepts of culture and human nature. It is not possible to differentiate between activities that result in similar traumatized human remains when those activities differ only in their motivation. The accurate use of taphonomy can indicate only cause of death and how bones are modified, not why. Correct analysis of human remains from Anasazi mass graves indicates death by blunt-force trauma, with perimortem and postmortem corpse mutilation. Reported bone breakage patterns have failed to stand up to scrutiny, and the fecal material containing human tissue is now considered likely to be from a scavenging coyote. Although cannibalism may have occurred among the Anasazi, particularly in starvation situations, it cannot be proven through methodology that fails to consider culture as one of its variables.

## Anasazi Warfare

However, acceptance of the idea that violence occurred among the Anasazi made it possible to understand the previously overlooked evidence of warfare. This includes recognition of its victims. A few individuals in formal burials have been found with stone projectile points embedded in their skeletons, or with their skulls crushed by blunt-force trauma. But burials of this type are rare. Most victims of Anasazi warfare are now known to be the individuals represented by the masses of fragmentary human remains found in mass graves.

We now know that warfare was a common aspect of life among the prehistoric Anasazi. By using recorded historic American Indian practices as probable examples, it has been possible to create a model of Anasazi warfare. The basic pattern of Anasazi warfare seems to have been one of constant low-level raiding between neighboring communities, periodically interrupted by large-scale battles and massacres. Individuals may have been ambushed as they worked in their fields or went for water. Small communities seem to have been massacred in large attacks. Quite often, these warring communities were related. Although a range of weapons may have been used, the most common weapon appears to have been the club. Large numbers of stone club heads (called *mauls*) are a regular feature of Anasazi ruins. Blows from these weapons are responsible for the intense bone breakage and battering. Slashing cuts from chipped stone blades sliced through internal organs and left cut marks on bones. Not only were victims killed but also, both at time of death and after death, their bodies were commonly mutilated. The dead were dismembered by individuals skilled at butchering game, and their bones were broken repeatedly by clubs.

The dead were left exposed for varying periods. Decay, the elements and scavenging animals further scattered the bodies. In most cases, it appears that the survivors of attacks and relatives of massacre victims returned when it was safe, some time after the event, to collect the portions of the dead still recoverable. In some cases, the bones may have been scraped clean with stone knives, a practice common to many Native American tribes. The collected, fragmentary remains were often placed in an abandoned structure or room that served as an expedient tomb. Sometimes they were buried. These secondary burials of broken and battered human remains constitute the mass graves found at the Anasazi sites.

Surprisingly, some of the strongest support for this model of Anasazi warfare is provided in Native American oral tradition through the one specific mass burial it documents, the mass burial at Polacca Wash (southwest of the Hopi Mesas in northern Arizona). Hopi oral tradition states that this is the grave of evil people from the Hopi village of Awatovi. Captured by other Hopi who raided their village, they were led to Polacca Wash, where

**BELOW** In this example of butchered Anasazi from Northern New Mexico, human bone from several individuals is scattered across the floor of a partially excavated room.

93

they were tortured and killed and their bodies mutilated. Later, their remains were gathered by relatives and buried at the spot. Excavations at Polacca Wash reveal the same range of human remains, exhibiting the same range of modifications, as those found at other Anasazi mass graves.

Additional supporting evidence of Anasazi warfare is the presence of defensive architecture at Anasazi sites. Wooden palisades surrounded some villages. Other communities built their settlements in defensible locations. At Mesa Verde in southwestern Colorado, communities were built within large rock shelters that could be reached only by climbing ladders or hand- and toeholds up the steep cliff face. The large Anasazi communal structures, known as *pueblos*, were built for group defense, with few outside windows or doors. Entrance was through narrow, easily defended openings in the walls and through openings in the roof reached by ladders that could be pulled up to deny access. In some areas, line-of-sight visual contact between communities enabled a degree of regional defensive cooperation. The Chaco System of large communal structures connected by straight prepared roadways may have been a regional defensive alliance.

The world of the Anasazi was a dangerous place. Known enemies could exist in the next village, and unknown enemies could appear and attack at any time. For the Anasazi, as in most cultures, warfare was an important social institution that helped protect the community and served to cement group social cohesiveness by generating intense in-group feeling and establishing an us-versus-them mentality within the population. Both the presence of the mass graves of butchered Anasazi and the large numbers in which they occur demonstrate how important this concept of community was to these people. For them, it literally was a matter of life and death.

**BELOW** Northern New Mexico contains extensive deposits of volcanic tuff. The Anasazi dug networks of rooms in this soft material. These residential areas, reached by ladder, were easy to defend against enemy attacks.

# A War Monument in Gaul

A battle took place about 2,250 years ago in northern Gaul at what is now Ribemont-sur-Ancre, 12 miles (20 kilometers) northeast of Amiens. A tribe of Belgae, called Ambians, defeated some Celts from Armorica (now Basse-Normandie, the Le Mans area). Although thousands of warriors were involved and hundreds were killed, history tells us nothing of the event; it is archaeology that has revealed this battle by reconstructing the monument that was set up by the victors – the only monumental trophy known from Celtic Europe.

The site itself extends for more than half a mile (one kilometer) in Picardy and comprises a big enclosure surrounded by a ditch, with human bones around a space that must have contained a palisade. It was first spotted from the air by Roger Agache, the great French aerial photography specialist. The original excavations here, in 1966, led to a belief that this was an immense Gallo-Roman sanctuary. Gallic iron weapons and human bones were found but were interpreted as the remains of graves disturbed by the Gallo-Roman builders. Then, as excavations progressed, a different interpretation arose, that of a Celtic sanctuary built in the third century BC.

However, in the 1980s, an exceptional assemblage of human bones was unearthed in the form of a cubic construction with daub walls made with arm and leg bones, in rows and interlaced, as well as a large charnel house. Because human bones are rare in Celtic sanctuaries but animal bones (mostly cattle) from sacrifices abound, this site was clearly unusual. The 20,000 human bones recovered must represent hundreds of victims, which meant that sacrifice was most unlikely.

Further analysis in the 1990s showed that the remains came from about 500 individuals, but they formed a group that was different from those encountered in the graves of the period. All these remains came from men – adults and young adults aged from about 15 to 40 – with not a single woman present. The men were also quite large, their average stature like that of modern Frenchmen, and robust, while their skeletons revealed no traces of illness or degeneration.

The remains from outside the enclosure, a rectangle measuring 16 by 5½ yards (15 by 5 meters), constituted an incredible tangle of weapons and pieces of skeleton – mostly long bones, with no skulls present at all. The numerous cervical vertebrae display traces of decapitations, mostly carried out with knives on dead bodies lying on the ground – a difficult task that can take half an hour because of the numerous muscles to be cut. These bones – from about 140 individuals, probably the defeated – had not been buried deliberately, but after the headless corpses had been

ABOVE Many of the vertebrae found at the site bear traces of butchery, and show precisely how the heads were severed from the bodies.

BELOW Precise measurements are taken from all of the thousands of bones unearthed at the site; at the same time, a detailed study is made of any traces of butchery or violence on them, such as sword cuts or spear thrusts.

exposed like human scarecrows for perhaps 200 years, gradually collapsing along with their support, sediment had accumulated on them naturally over a long period. The heads of the defeated were probably carried off as trophies by the battle's victors, attached to their horses.

More than 5,000 weapons were found, mostly fragments of spears, javelins and shields, but few scabbards and swords. In short, they reflect what can be collected from a battlefield – used, lost or broken weapons, or items like the shields that are thrown aside when fleeing. But an individual and prestigious weapon like the sword would have been kept by its owner or carried off as booty by the owner's killer.

The condition of some of the weapons indicates that this material was probably originally at some height, on a platform, and then fell to the ground when it collapsed. As so many hundreds of corpses and weapons were involved, as well as horses and chariots, the monument obviously was erected not far from the battle and perhaps on the battlefield itself.

About 15 gold coins found at the site place the battle around 260 BC and indicate that they probably belonged to the Lexovii, a people who lived in Basse-Normandie at that time. A gold torc confirms accounts by classical authors that Celts wore these in combat. Pollen and seeds unearthed at the site indicate that the battle took place at the end of summer.

The bone assemblages inside the enclosure, along the ditch, are more varied, but the most remarkable are the above-mentioned cubic constructions of long bones; these were about 1.6 yards (1.5 meters) square and nearly 2 feet (60 centimeters) high. In their center was a hole filled with crushed and burnt fragments of human bone. Six of these structures have been found so far, with about 2,000 bones from 300 individuals; horse bones are mixed with the human remains.

The remains inside the enclosure seem to have been treated in the same way as those found in the normal graves of the period, so this is likely the collective grave of the victors, as is confirmed by the presence of horses, which were accorded great importance and often buried with the dead. It is thought most likely that the victors were Belgae who arrived in this region of northern Gaul in the mid-third century. Classical writers inform us that warriors' bodies were left exposed for birds to pick their bones clean; only then could the bodies join the gods.

The area became a place of cult for two or three centuries, with offerings of weapons and banquets occurring on its periphery. When the Romans arrived in the region, the remains of the monument were destroyed, and a Roman temple was erected inside this place of honor in about 30 BC.

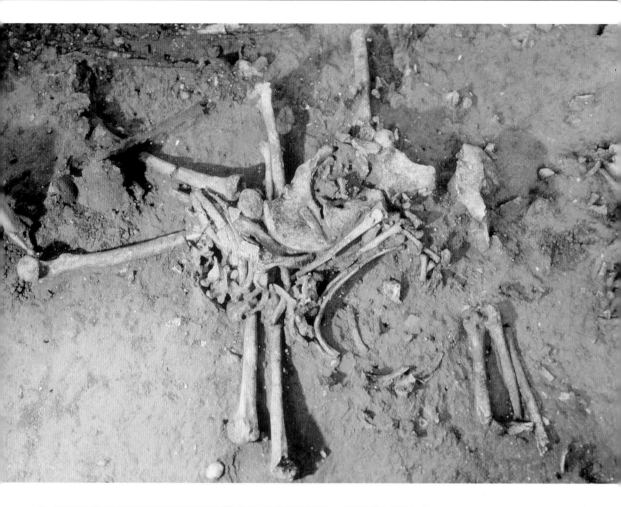

**ABOVE** Dirt is carefully and gently brushed from one of the tens of thousands of human bones unearthed at Ribemont since excavations began in the 1980s.

**LEFT** A selection of some of the thousands of metal weapons found at the site, including spearpoints and arrowheads.

# Windeby Girl:
## An Iron Age Bog Body

I n May 1952, peat-cutters working at a small bog near Windeby in the German state of Schleswig-Holstein noticed parts of human limbs on their conveyor belt. Investigation where the peat had been cut revealed that these had been shaved off a preserved body buried in the bog. Police were called, but it soon became apparent that the body was not that of a recent murder victim but rather was an ancient corpse. The police were replaced by archaeologists, who investigated the peat deposit further.

The archaeological research revealed not one but two prehistoric corpses lying 5½ yards (5 meters) apart. One was an elderly man. Only his skin remained, and his body has been pressed flat. The other was the well-preserved body of a girl. She lay on her back with one arm outstretched. The girl was naked except for a leather collar and a strip of cloth over her eyes. The peat workers and the archaeologists had discovered Windeby Man and Windeby Girl, two of the many "bog bodies" that have been found across northern Europe.

## Bog Bodies in Northern Europe

The term "bog bodies" refers to the hundreds of corpses discovered in the peat bogs of northwestern Europe whose soft tissues – skin, internal organs, hair – have been preserved along with the bones and teeth. Peat bogs are acidic wetlands, where the rate at which plants grow and die outpaces their rate of decay. The undecayed remains accumulate as peat. For centuries, peat was used primarily as a fuel in home hearths and craft production. Recently, it has become popular for enriching garden soil. Until recently it was cut by hand, but now machines shave away the peat in layers and load it onto conveyor belts for packing.

Why does peat preserve human remains so well? Waterlogged peat contains little oxygen and, as a result, few of the microorganisms that promote decay. Of course, it is necessary for the body to become submerged in this oxygen-poor environment quickly and not lie exposed to air and microorganisms. The acidity of the bog also minimizes the number of microorganisms. It also

BELOW Bogs and small ponds were sacred places in northern Europe during the Iron Age – between 300 BC and AD 200 – when deities were appeased with ritual sacrifices of people and animals, as well as gold and silver.

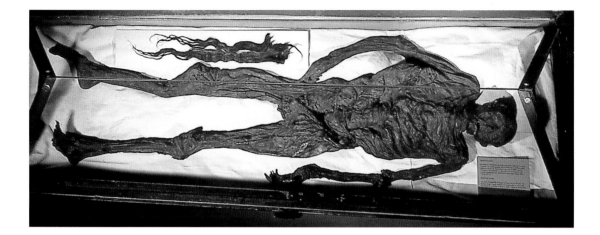

appears that the chemistry of the bog does more than simply prevent decay; it actually promotes preservation. Sphagnum moss, the primary plant constituent of peat bogs, produces a substance called *sphagnan*. When sphagnan is released from dead moss and comes into contact with a body, it has two effects. First, it extracts calcium from the body, which gives bacteria less to feed on. Second, it has a tanning effect on the skin and simultaneously binds nitrogen, making it also unavailable to bacteria. For this reason, the skin of bog bodies is usually dark brown and the bones are often decalcified, leading to their deformation and possibly even disappearance.

## The Classic Bog Bodies

The first recorded bog bodies were found in the eighteenth century, but not until late in the nineteenth century was it realized that these bodies came from prehistoric times. Archaeologists observed the layers in the bogs, the proximity of prehistoric settlements and artifacts associated with the bodies. The evidence pointed toward an Iron Age date for most of the bodies, which across northern Europe means the few centuries before and after Christ, around 2,000 years ago.

Horrified peat-cutters in the Bourtangermoor in the northern Netherlands came upon Yde Girl in 1897. Her most distinctive feature was her red hair, more than 8 inches (20 centimeters) long, but it was on the left side of her head only; the hair on the right side had been shaved off. Sixteen-year-old Yde Girl had been strangled by a woolen belt that was still wrapped round her neck, although a stab wound at the base of the neck probably contributed to her death.

In 1938, the body of a woman was found at Bjaeldskov Valley near Silkeborg in Denmark. Her hair, over 1 yard (1 meter) long, had been braided and tied in a bun, and her body had been wrapped in a sheepskin cape with a leather cloak around her legs. Subsequent investigation showed her to have been 25 to 30 years old. Around her neck was a deep

**ABOVE** Haraldskaer Woman was found in the early 19th century in Denmark. Originally thought to be the remains of the legendary Norwegian queen Gunhild, radiocarbon dating showed that she was the victim of an Iron Age sacrifice.

**BELOW** Human hair is usually well preserved in the acid bog sediments, so bog bodies provide evidence for Iron Age hairstyles, which often involved elaborate braiding and knotting.

groove, indicating that she had been hanged or strangled, probably with the leather rope found nearby.

On May 8, 1950, another bog body was found in the Bjaeldskov Valley, about 77 yards (70 meters) away from that of the woman found 12 years earlier. This body was marvelously preserved, and it was carefully excavated by lifting it together with the surrounding block of peat, although the whole thing was so heavy that one of the workers collapsed and died of a heart attack. The block of peat with its body was sent to the National Museum in Copenhagen for analysis. The bog where the man was found is called Tollund, and he became known as Tollund Man.

Like the woman found in 1938, Tollund Man died from hanging. The braided leather rope still around his neck left no doubt. He wore a pointed leather cap held under his neck by a thong, and around his waist was a leather belt. Otherwise, he was naked. He was lying in a crouched position with an eerily peaceful expression on his face. Tollund Man was probably 40 to 50 years old. Analysis of his stomach contents showed that his last meal had been a gruel of barley, wheat and flax mixed with many weeds.

Two years later, peat-cutters at Grauballe, about 11 miles (18 kilometers) from Tollund, came upon another body. Grauballe Man lay on his chest, naked, with his left leg extended and his right arm and leg bent. Lessons learned at Tollund were applied at Grauballe, and the block of peat with the body was taken

**RIGHT** Tollund Man is probably the most famous of the bog bodies. He is shown here as he was found – lying on his side with his conical leather cap and the cord that strangled him around his neck.

to the Museum of Prehistory at Aarhus. The cause of death was obvious: a slash ran across the front of his neck from ear to ear. Seeds of 63 different plant species were identified in Grauballe Man's digestive tract.

Carbon-14 dating of Tollund Man and Grauballe Man confirmed their Iron Age date. A sample of Tollund Man's tissue has been dated to 220 +/- 55 BC, while Grauballe Man has been dated to 265 +/- 40 BC. The woman found in 1938 has now been dated to about 210 BC. Other bog bodies across northern Europe have been dated to the final centuries BC and the first centuries AD.

In August 1984, a worker at a peat-shredding mill near Lindow, a town near Manchester in England, picked up what looked like a piece of wood and threw it toward a coworker. When it hit the ground, out popped a human foot. The local archaeologist examined the peat cutting where it had been found and identified the remainder of the body, the torso of a male, about 25 years old, with a beard, mustache and short hair. On his arm was a band of fox fur; otherwise, he was naked. In the laboratory, the peat was cleaned away, and it became clear that Lindow Man had not only been strangled but also had his throat cut, and he had suffered violent skull injuries as well. Someone really wanted him to die, sometime in the first century AD.

ABOVE Lindow Man's head, emerging from the peat, shows his beard (very rare on bog bodies) and trimmed hair. Closer examination revealed that he had been hit with a blunt object before being strangled.

## Iron Age Society

The society that produced the bog bodies was a mysterious world of ordinary farmers, skilled craftsmen, bands of warriors and powerful ritual leaders. Archaeologists refer to this period as the Iron Age because it was then that iron metallurgy came into use for tools and weapons. The Iron Age societies of northern Europe were described by the historian Tacitus and encountered by the Romans during their forays into Gaul, Germany and Britain. The communities whose members wound up buried in bogs were beyond the view of the Romans, so our primary knowledge about them comes from archaeological finds.

In Denmark and northern Germany, the Iron Age people lived in long houses, between 16 and 33 yards (15 and 30 meters) long, with two rows of posts running down the center supporting thatched roofs. These houses were multipurpose structures, with living space for people at one end and stalls for animals at the other. In the British Isles, however, Iron Age houses were generally round, about 11 yards (10 meters) across, with a central

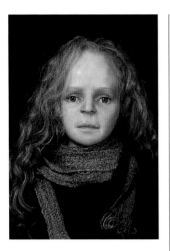

**ABOVE** A medical artist has reconstructed the appearance of Yde Girl, whose contorted and desiccated corpse in the Drents Museum in Assen, the Netherlands, had to be meticulously studied to determine her original facial structure.

**BELOW** The disturbing image of Windeby Girl, blindfolded with her head partly shaved, inspired the 1975 poem "Punishment" by the Irish poet and Nobel laureate Seamus Heaney, in which he calls her a victim of "tribal, intimate revenge."

hearth. Animals were kept in outbuildings, while grain was stored in underground silos.

Such farmsteads were generally grouped into hamlets. Some sites were the seats of chiefs who commanded the loyalty of subordinates and commoners. At their service were craftsmen who produced finely made objects of bronze, gold and silver. The chiefs also led groups of warriors that conducted raids on rival communities.

The Iron Age world also had a mysterious spiritual side. We know little about the people's beliefs, but it is clear that watery places held considerable significance. Ponds and bogs across northern Europe have yielded ritual deposits of pottery vessels, metal ornaments, wooden idols, weapons, animals and war booty. These were not accidental losses but rather deliberate acts of sacrifice to powerful forces. The human corpses found in the bogs were apparently also sacrifices to these forces. Although some archaeologists have proposed that they were criminals or outcasts, the widespread use of bogs and ponds for sacrifice makes it more likely that their deaths served some spiritual purpose.

## The Death of Windeby Girl

Thus we return to Windeby Girl, sacrificed in a bog about 2,000 years ago. After her blond hair was shaved on the left side of her head, the teenager was led to the bog blindfolded and drowned, probably in very shallow water. A large stone and some branches were placed on her to hold her down. How was she chosen for such a brutal fate? What did her killers hope to accomplish? What powerful forces were they trying to appease? We will never know the answers, but Windeby Girl and the other bog bodies offer a stark glimpse of a deadly Iron Age ritual.

# Who – or What – Killed
## Tutankhamen?

I n 1922, Howard Carter discovered Tutankhamen, Egypt's "boy pharaoh," lying secure in his tomb in the Valley of the Kings. His mummified body rested in a sealed nest of three coffins, protected by a quartzite sarcophagus and surrounded by golden treasures beyond Carter's wildest imaginings. But Tutankhamen's tomb was in some respects disappointing: It held few written records, and it failed to tell the story of the dead king's life. We must therefore rely on the evidence gleaned from Tutankhamen's own body if we wish to understand the circumstances surrounding his premature death.

## The Physical Evidence

Tutankhamen was approximately 18 years old when he flew to meet his gods. He had already accomplished a great deal, having ruled as pharaoh for almost a decade. He had married, and the miniature golden coffins of his two stillborn daughters were included in his tomb. But, as a dynastic Egyptian male who had survived the perils of birth and infancy, he could reasonably have expected to live to 40 or even 45 years of age. Indeed, as a pharaoh, he had a good chance of living far longer; Egypt's elite benefited from good food and hygienic living conditions, and were spared the physical dangers that shortened so many working-class lives. Some pharaohs managed to beat the statistics and live well into their nineties. What, then, had brought about Tutankhamen's untimely death?

Tutankhamen was discovered at a time when forensic archaeology was in its infancy. Although he treated the king with a brusque respect, Carter had little interest in mummies, and he regarded Tutankhamen as simply another artifact to be processed. The thick resins used in his funeral ceremonies had caused the king's bandages to adhere to his inner coffin. The anatomist Professor Douglas Derry was given the delicate task of detaching Tutankhamen from his coffin.

In 1925, an autopsy was performed in the cramped and dimly lit tomb of Seti II. Derry found the king's bandages almost completely carbonized. While it was possible that this had been caused by a natural combustion within the mummy wrappings, Carter's use of heat to separate the trio of coffins may well have been a contributing factor. Derry removed the decayed bandages and dismembered the king. Finally, the golden mummy mask was removed with hot knives. Tutankhamen was revealed as a young man, clean-shaven, with a flat nose and the pierced ears favored by Egyptian boys. The end of the autopsy saw Tutankhamen lying, unwrapped and in pieces, on a tray of sand in his inner coffin.

**ABOVE** Included in Tutankhamen's tomb were two miniature coffins designed to hold the mummified fetuses of two female children. These children are unnamed, but many archaeologists assume that they are the daughters of the boy-king.

It had been intended that the king be X-rayed before the unwrapping, but the radiographer died on his way to the tomb. Two sets of X-rays were eventually taken – by Professor R. G. Harrison (in 1968) and by Dr. James E. Harris (in 1978). Today, the body lies sealed in its coffin in the Valley of the Kings. It is inaccessible to all.

## The Suspects

We know that Tutankhamen lived in difficult times. His father, the heretic Akhenaten, had abandoned the old gods, turned his back on diplomacy and retreated to an isolated new city, Amarna. The nine-year-old Tutankhamen inherited a bankrupt throne and a discontented country on the verge of anarchy. Fortunately, he also inherited a cabinet of experienced ministers who could guide his uncertain steps. Chief among these was Ay, a wise and respected politician who had already served under three pharaohs. Also prominent were the respected veteran General Horemheb and the queen, Ankhesenamen, daughter of Akhenaten and half-sister of her husband, Tutankhamen.

**ABOVE** Within the confined, hot and humid burial chamber of Tutankhamen's tomb, Howard Carter was forced to construct makeshift lifting equipment in order to open the successive gold coffins.

## The History

Egypt's kings liked to think themselves invincible. Appointed by the gods and themselves semidivine, they could not be challenged by mortal men. We know that regicide did occur – at least one pharaoh, Ramses III, was assassinated in a conspiracy

**BELOW** The mummified fetuses, discovered by Howard Carter, still lie wrapped in the wool he used to protect them. Egyptologists hope they carry the genetic material of the late-18th-Dynasty royal family.

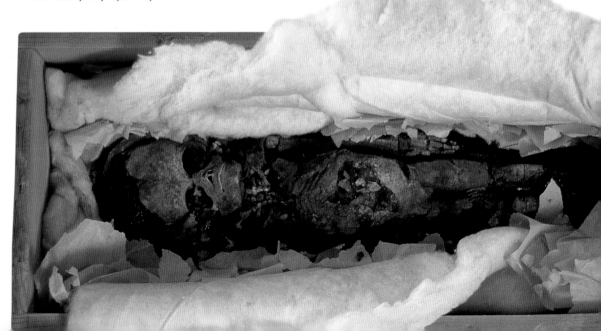

hatched in the royal harem – but so strong was the official propaganda, so abhorrent that a commoner could terminate a royal life, that such cases went unrecorded and unmentioned. Tutankhamen's death, whatever its cause, was treated by his people as an unfortunate but unsuspicious tragedy.

The death took everyone by surprise. As there was no male heir and no royal tomb prepared, Ay took charge. Burying the pharaoh in his own cramped and hastily decorated tomb, Ay was crowned king of Egypt. But Ay was already an old man, and within four years he too was resting in a tomb in the royal necropolis. Now it was General Horemheb's turn to become pharaoh.

Not everyone was happy with the new regime, and at least one person was driven to take unprecedented action. A letter addressed by a widowed queen of Egypt to the king of the Hittites, an enemy of long standing, asked that a Hittite prince be sent to marry her and become pharaoh, "for I cannot marry one of my servants." Unfortunately, the name of the letter writer is not preserved but, as the letter is approximately contemporary with Tutankhamen's death, we can guess that it was signed by Ankhesenamen. It is always possible that the letter was a trap to snare the Hittites; if this was the case, it succeeded brilliantly, as the prince who set off to become pharaoh was murdered as he crossed the border, and Egyptian-Hittite relations plummeted to an all-time low. If the letter was indeed a genuine call for help, it failed. Ankhesenamen did not retain her position as queen, and this is the last that we see of her.

**ABOVE** The golden mask which rested on the mummified body of Tutankhamen is perhaps the most fabulous and famous find to come from the ancient world. Today it is the centerpiece of the display in Cairo Museum.

## Suspicions

Tutankhamen lived surrounded by highly ambitious courtiers. He died unexpectedly, well before his time, and his queen disappeared soon after his death. Two of his courtiers succeeded him on the throne. In 1923, Arthur C. Mace voiced the suspicion that Tutankhamen was murdered, laying the blame at Ay's door. Others have agreed with his diagnosis of murder but have cast Horemheb as the villain. Suspicion is one thing, however, and proof another. As it is not possible to reexamine Tutankhamen's body for traces of poison or violence (Egyptian authorities will not allow anyone to unseal the coffin), and as we have absolutely no written evidence to guide our investigations, all modern reconstructions are through necessity based on the work done by professors Derry and Harrison and by Dr. Harris.

## Proof

Attention has focused primarily on Tutankhamen's head. Analysis of the 1960s skull X-ray shows a detached piece of bone within the skull; experts, however, agreed that this should be eliminated from the murder investigation as it is the result of post-mortem damage sustained in the embalmer's workshed and so could not have contributed to Tutankhamen's death. Of considerably more interest is a region of thickening or darkness just where the head joins the neck at the base of the skull. Skulls vary in their bone density, and Harrison, professor of anatomy at Liverpool University, considered this area of thickening well within the accepted normal range. While all subsequent investigators have agreed with his diagnosis, there remains the possibility that the thickening is not normal, that it indicates the site of a hemorrhage caused by a forceful blow to the back of the head. Such a blow would have been strong enough to cause death, and it is in a position that makes it unlikely to have been caused by a simple fall.

Recently, Dr. Bob Brier of Long Island University has highlighted an area of clouding or fogging in the region of the skull thickening. This may be entirely natural or the result of an error in the X-ray technique; it may also represent a calcified membrane formed over a blood clot. This in turn would suggest that Tutankhamen, having received a stunning blow to the back of the head, lived long enough for at least partial healing to occur, a process that would take at least two months, if not far longer. This means that the blow, if it ever occurred, may not have been the direct cause of his death.

While expert attention has focused on Tutankhamen's head, the rest of his body has been largely ignored. Almost a century ago, Professor Derry noted substantial damage to the king's chest and rib cage, sections of which were already missing before the body was mummified. It may be

**BELOW** Many artifacts recovered from Tutankhamen's tomb show the king with his wife and half-sister Ankhesenamen. This confirms the genetically confusing practice of brother-sister marriages within the 18th-Dynasty royal family.

that this represents post-mortem damage sustained in the embalming shed. But, although we know that Egyptian body parts frequently went missing after death, this seems unlikely; and we are perhaps entitled to assume that, no matter how careless they were with commoners, the undertakers took the utmost care when preserving their dead pharaohs. The pharaoh would have been embalmed in his own dedicated embalming house soon after his death and the embalmers would have been very careful with his body. It therefore seems that Tutankhamen's whole body was badly damaged at the time of his death. This in turn suggests that he may have died as the result of an accident rather than an illness or a simple blow to the head.

Fatal accidents were a sad fact of life in ancient Egypt. Men drowned in the Nile or were eaten by crocodiles; they perished in mines and quarries, in foundries and on building sites, and were even crushed by falling rocks when robbing graves. The medical papyri suggested ways that doctors might treat severe blows to the head and body, but medical techniques were severely limited and, in most cases, there was no hope of recovery. We know that Tutankhamen rode a golden chariot; it was buried with him in his tomb. All the New Kingdom pharaohs were expected to be daring huntsmen; victory in the chase, the slaughter of wild lions and bulls, was seen as a sure and certain sign of divinely inspired kingship. Could we be looking at the unfortunate result of a high-speed chariot crash on the hunting field?

## The Verdict

Tutankhamen died young and of no obvious illness – although, as it has not proved possible to examine his tissues using modern medical techniques, we cannot rule out the possibility that he died of a simple yet deadly disease such as measles, diarrhea or parasitic infection. We do know that he sustained severe damage to his chest, and he may also have suffered a crushing blow to the head. The damage to both head and chest may have been deliberate or accidental; it may even have been caused after death but, had it occurred during his lifetime, it almost certainly would have killed him. Murder remains a possibility, with Tutankhamen attacked from behind with a club or stout stick or perhaps encouraged to crash in his chariot. If he were murdered, both Ay and Horemheb must remain chief suspects, as both undoubtedly benefited from his early demise. Ankhesenamen, who lost by her husband's death, is unlikely to have colluded with his enemies and cannot, therefore, be considered a serious suspect.

The evidence for murder is, however, by no means overwhelming. The damage to the chest allows us to reconstruct with equal, or more, validity a tragedy on the hunting field wherein the young king, determined to prove his strength and ability, was trampled by horses' hooves or crushed beneath the frame of his chariot.

**ABOVE** The mummy of Tutankhamen, stripped of all its bandages, as unwrapped by Carter's archaeological team. Today the boy-king still rests in his sarcophagus in the Valley of the Kings.

# The Cap Blanc Lady

The limestone rock shelter of Cap Blanc, near Laussel, northeast of Les Eyzies in France's Dordogne region, is well known to the world of prehistory as the site of one of the finest sculptured friezes to survive the last Ice Age, the first to be unearthed, and currently the best to remain open to the public. Its figures of horses, bison and deer, albeit much damaged at the time of their discovery by Dr. Gaston Lalanne of Bordeaux in 1909, remain a moving and powerful ensemble. Lalanne dug here and unearthed a fine collection of typical Magdalenian – about 15,000 years old – stone, bone and antler tools, including harpoons, and a number of large stone implements that had clearly been used to produce the parietal bas-relief and haut-relief sculptures that his crude excavations brought to light on the back wall.

In 1911, further digging in front of the shelter for the purpose of erecting a small construction to enclose and protect the frieze and for lowering the floor level to make the art more visible to visitors led to the discovery of a human skull. Work was suspended and prehistorians Louis Capitan and Denis Peyrony were asked to extract the skeleton, a task that took them three days.

The Cap Blanc skeleton is of tremendous importance – not only a relatively intact inhumation from the late Ice Age but also one of the very few found in close proximity to parietal art of the period. Indeed, the body's location directly in front of the central part of the shelter's sculptured frieze can really only be compared with that of the double paleolithic inhumation of an adult woman buried with her arm around a 17-year-old male dwarf in front of the engraved block at the Riparo Romito, Italy (see p. 148). It was suggested by the excavators that the Cap Blanc burial may even be that of the original sculptor (or one of them), and this is unquestionably a possibility; certainly the location of the inhumation indicates a person with a strong link to the site.

## Conflicting Reports

In France, the excavation of the skeleton in 1911 led to a brief publication that discussed primarily the two skeletons unearthed at La Ferrassie by the same excavators. They gave few details about the Cap Blanc find, stating only that the skeleton lay at the bottom of the archaeological deposit, 2½ yards (2.3 meters) from the frieze and 2 feet (60 centimeters) below the hooves of the central horse. It had been buried amid stones, with three fairly big stones placed above it, one of them on its head and others at its feet. It had been placed on its left side, arms and legs flexed, occupying a space of only 3 feet by 2 feet (1 meter by 60 centimeters), immediately below a Magdalenian hearth.

It is curious that early reports of the Cap Blanc skeleton claimed that it was of a male aged about 25, whereas examination by physical anthropologists eventually established that it was of a young adult female.

A recent examination of the Field Museum's archive on the case made it possible to reconstruct much of the story. The earliest document in the archive is a letter, dated January 24, 1911, to Mr. J. Grimaud, the site's owner, from the president

ABOVE Two of the horses sculptured in haut-relief and bas-relief, facing right, on the back wall of the limestone rock-shelter of Cap Blanc. They are 6 feet (1.9 meters) (left) and 4½ feet (1.4 meters) (right) in length.

of the Société des Antiquaires de l'Ouest in Poitiers, acknowledging receipt of a report on the rock shelters of Laussel (i.e., Cap Blanc) together with photos and five boxes, one containing reindeer teeth and bones and the other four containing flint tools. A letter, dated August 5, 1911, from Paul Léon, at the Ministère de l'Instruction Publique et des Beaux-Arts in Paris, thanks Mr. Grimaud for reporting the discovery of the skeleton and states that he will ask Peyrony to take appropriate measures to preserve it. Peyrony himself (the Membre correspondant de la commission des monuments historiques aux Eyzies) writes on August 8 that the Minister has asked him to verify the authenticity of the Laussel skeleton, make all necessary scientific observations, and supervise the excavation. He therefore went to the site that very morning and examined the find in the presence of Grimaud's guard, Veyret. The remains were indeed authentic.

Only two days later, Grimaud received a letter from Dr. Capitan, professor at the Collège de France, dated August 10, which is a key document for the site. The letter contains a sketch of the location of the bones and reports that they are 2½ yards (2.3 meters) from the big horse and around 27 inches (70 centimeters) below its muzzle. They occupy a kind of pit, 20 inches (50 centimeters) deep, and the skull was unfortunately broken by a blow from a workman's pickaxe. Capitan insists, rightly, that the excavation be carried out by experienced and qualified people and suggests himself and Peyrony for the task, as they have just unearthed the two older skeletons from La Ferrassie. To make matters clear, he proposes that the excavators produce the scientific report, while any finds would belong to Grimaud. In the meantime, the skeleton has been covered with stones and planks for its protection.

A new letter from Capitan, dated August 28, reports that the skeleton

ABOVE Television lighting helps to bring out the relief of the magnificent frieze at Cap Blanc, in particular the great central horse, 7 feet (2.2 meters) in length. As with all Ice Age bas-reliefs known so far, it is thought that the frieze was originally painted too.

has been removed in its entirety in a number of blocks of earth, and it will now be possible to excavate the bones properly and carefully, once Peyrony has transported them to Paris by rail, probably in September or October. For the present, these blocks are in Peyrony's care, and he will dry them out slowly. Most important is a brief sentence, stating that "All we found with the skeleton was a shapeless fragment, probably of ivory." This is indeed a small ivory point measuring 0.6 by 3 by 0.4 inches (16 by 74 by 10 millimeters), which is kept at the Field Museum, having been sold along with the skeleton. It is described as "several thin laminae glued together along with bits of matrix and partially reconstructed or plastered over with some sort of filling material." According to its original display case label, this point was "found over the abdominal cavity of this individual," and "the weapon may have been the cause of death."

This is certainly the theory that was promoted by Henry Field, the eventual acquirer of the skeleton for the museum. He claimed in a 1927 article that the skeleton died a natural death, yet also noted:

*A small ivory harpoon-point found lying just above the abdomen may give a possible clue to the cause of his death. This weapon may have caused blood poisoning which resulted in death. It has been suggested tentatively that the young man [sic] felt death approaching and returned to the rock-shelter, as he desired to die before the masterpiece he had helped to create... It is not plausible that some one who had nothing to do with the sculpture should have been allowed to desecrate the sanctuary unless he had assisted in the work or, at any rate, was directly connected with it.*

In Field's memoirs, his speculations were even more romantic: "Why had she been buried beneath the frieze of horses? Was she killed by her lover's ivory lance point? Was it by another Cro-Magnon girl? Was her brother avenging the family's honor? Was she killed in battle? Why was she buried in the sanctuary? Was she the daughter of the sculptor–high priest? There was no real evidence, except that death probably resulted from blood poisoning."

No source is given for the theory that the ivory point was the cause of death or the claim that it was found above the abdomen – perhaps this was merely Mr. Grimaud's opinion – but nevertheless it is baffling that such a potentially important object was completely omitted from the published report by Capitan and Peyrony. Indeed, were it not for this casual mention in Capitan's letter, there would be absolutely no guarantee that the point had any connection with the Cap Blanc skeleton. Yet ivory is not common in Magdalenian contexts in southwest France, let alone ivory points that may be a cause of death. In this connection, it is worth noting that the only clear evidence we have of violence inflicted on humans during the last Ice Age consists of a probable flint arrowhead embedded in the pelvis of an adult woman from San Teodoro Cave, Sicily, and an arrowhead in the vertebra of a child from the Grotte des Enfants at Balzi Rossi, Italy.

A letter to Grimaud from Peyrony, dated August 31, 1911, notes that "we have been able to lift the whole thing in a pretty good state. The whole skeleton will be able to be reconstructed and will be a very good study piece. I have conserved it in Les Eyzies, as Mr Capitan was not able to take it. I will carry it to Paris next October." However, it is clear that Capitan had major problems in getting the skeleton dealt with in Paris. Letters from him complain of the difficulty in finding someone qualified and with sufficient time available to prepare the bones for casting and display. It is also interesting to learn that there were plans afoot to have a cast made and placed in the shelter; in fact, for some reason this was never done, and instead a miscellaneous collection of casts of other bones was put together for this purpose. In a letter dated July 29, 1913, Capitan tells Grimaud that an artist will be sent to carry out this assignment. A letter from Grimaud in 1924 notes that "in accordance with the Ministère des Beaux Arts, I have had a modern skeleton set in place at the foot of the sculptures, in place of the real skeleton."

Nevertheless, the original skeleton was eventually extracted from its sediments by J. Papoint of the Laboratoire de Paléontologie at the Musée National d'Histoire Naturelle under the direction of Marcellin Boule (director of the museum) and of Capitan. A letter from Papoint, dated February 27, 1915, records the state of the bones:

*You will find the skull in the wooden box. It is in two pieces. It was impossible for me to reconstruct it because of the deformation caused by fossilisation. I left in the same block the upper and lower jaws as well as the seven cervical vertebrae which I extracted as well as I could. There are two upper incisors*

BELOW The cast of the Cap Blanc lady, restored to her original resting place in front of the center of the carved frieze on July 14, 2001.

*that I put to one side, since I could not fit them in their sockets. These two skull pieces are very fragile and need to be unpacked with care. The dorsal and lumbar vertebrae are all present. The ribs are incomplete. All the limb bones are in good condition. A few fragments of the shoulder-blades and pelvis bones are missing. This is due to the fragility of certain parts of these bones. A few phalanges are missing from the hands and feet.*

## The Sale of the Bones

By early 1915, the Cap Blanc skeleton had been restored to its owner, Mr. Grimaud. It then disappeared from view until the start of his attempt to sell it to an American museum nine years later. According to Henry Field, "in 1916 M. Grimaud, having made no money out of the discoveries on his property, decided to reclaim his anticipated profit, and during the stress of war conditions was able to ship the skeleton to New York." In his later memoirs, he added that "the skeleton was said to have been smuggled out of France during World War I in a coffin as an American soldier with the necessary papers forged." Yet documentation available at the Field Museum provides no real clue as to why Grimaud decided to send it to America, or why he apparently waited a further eight years before trying to sell it. His initial choice was the American Museum of Natural History in New York, but, to cut a long story short, his protracted negotiations, via American lawyers in Paris, eventually came to nothing, in part because of his huge asking price ($12,000, equivalent to about $250,000 today). Finally, after steadily dropping his price, he sold it to Chicago's Field Museum for a much lower amount. According to Field's memoirs, a representative of the museum was sent to Mr. Grimaud "with twenty-five thousand-franc bills (the equivalent of a thousand dollars) in one hand and a receipt ready for signature in the other." He continues, "Some days later a cable came from Paris saying that the Cap-Blanc skeleton was ours. I hurried to New York and in the basement of the Museum of Natural History packed her very carefully in cotton wool and carried her in a suitcase to a compartment on the Twentieth Century [train]. We had a very uneventful night together."

With the benefit of hindsight, Field's memoirs claim that, as he laid out the bones in Chicago, "the pelvic girdle was definitely feminine" – yet, as noted above, his article of 1927 still saw the skeleton as a young man! The skeleton in its new case was first displayed prominently just inside the museum's main entrance.

**BELOW** The sketch of the Cap Blanc lady's resting place, in a letter sent by Louis Capitan on August 10, 1911, to the site's owner, Mr. Grimaud.

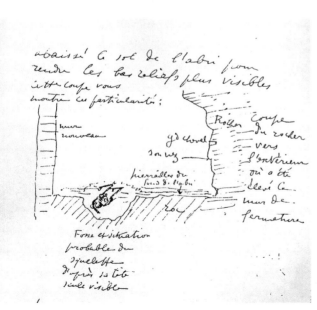

It was introduced to the media as "the only prehistoric skeleton in the United States" [sic], and so became front-page news. The first day, 22,000 visitors came to see for themselves. "At noon, the crowd was so dense around her that the captain of the guard...notified the director that two guards must be placed there to keep the people moving and orderly....[N]othing like this had happened before in the Field Museum.... This was the first exhibit in the new building to capture the public and press imagination."

In 1932, the skeleton was withdrawn from exhibition so that the skull could be restored by T. Ito under the direction of Gerhardt von Bonin of the Department of Anatomy at the University of Illinois. According to von Bonin:

**ABOVE** The excavation of the Cap Blanc skeleton by Louis Capitan (left foreground) and Denis Peyrony (right foreground) in August 1911. The carved frieze is to the right.

> *When the skeleton arrived at the Museum, it was in an almost perfectly clean condition, only a few bones being still embedded in a matrix of somewhat gritty, loam-like matter. The long bones were almost all perfectly preserved. The pelvic and the shoulder girdle were somewhat damaged, particularly in the pubic region and the scapula. The vertebral column appeared to be complete, the vertebrae were for the most part still held together by adhering soil. Twelve left and ten right ribs were found, and a rather decayed square piece of bone, apparently all that was left from the* **manubrium sterni**. *The cervical column was firmly attached to the lower jaw and a part of the upper jaw. The skull was broken into a number of fragments. The bones are of a brownish colour, darker in some spots and lighter in others. They are firm enough to be handled conveniently, yet somewhat brittle. In some spots, dental cement had been put on the bones in order to prevent them from crumbling.*

Von Bonin's conclusion, after a full anatomical study, was that these were the remains of a young woman, about 5 feet, 1 inch (156 centimeters) tall and about 20 years of age. In an exhibition case next to the skeleton, the museum installed a life-size diorama of the Cap Blanc rock shelter, modeled by Frederick Blaschke. As the only complete European paleolithic skeleton on exhibition in an American museum, the Cap Blanc woman was seen by several million visitors in her first decade in Chicago alone. But the story does have a happy ending of sorts. Thanks to the generosity of a private sponsor, a complete cast of the Cap Blanc lady – and of her ivory point – was recently made, and on July 14, 2001, the cast was installed in its rightful place beneath the central frieze in France.

# *Batavia's* Graveyard

RIGHT Popular accounts of the *Batavia* mutiny were produced within a few years. This illustration of the massacres on Beacon Island comes from Jan Jansz' *Ongeluckige Voyagie van't Schip Batavia*, published in 1647.

BELOW Diagram showing the relative positions of skeletons in the mass grave. The black deposit was removed and excavated later in the laboratory. As well as small bones, it contained unerupted teeth from an infant about nine months old.

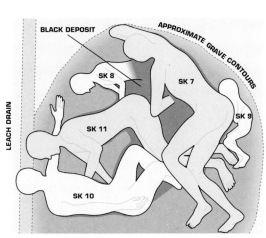

BLACK DEPOSIT
APPROXIMATE GRAVE CONTOURS
SK 8
SK 7
SK 9
LEACH DRAIN
SK 11
SK 10
EDGE OF EXCAVATION TRENCH

On June 4, 1629, the United Dutch East India Company's ship, *Batavia*, struck a reef in the Houtman Abrolhos Islands, off Geraldton, Western Australia. She was bound for Batavia (today's Jakarta). On board were over 300 people, including passengers and about 100 soldiers, as well as the crew. In the chaos following the wreck, about 40 people drowned, while the remainder landed on tiny Beacon Island.

Francisco Pelsaert, the Commander, and Captain Adriaen Jacobsz decided to sail the two small boats north to Batavia for help, leaving about 225 desperate people stranded. But there had been trouble on the voyage and Jeronimus Cornelisz, the undermerchant, was already planning a mutiny. Now he exerted his authority as a company official to control the situation on the islands. The plan was to capture the rescue ships when they arrived. To that end, he and his followers conspired to kill most of the remaining survivors.

The murders began on July 3, 1629. At first the killings were stealthy – the mutineers lured people from their tents at night and stabbed them, or took them fishing on rafts and drowned them. The most bloodthirsty killing was perhaps the murder of the Predicant's family, while he and his eldest daughter were dining with the mutineers. The soldiers loyal to the company presented the greatest threat. Accordingly, Cornelisz sent most of them to explore West Wallabi Island and stranded them there. He declared himself "Governor" and began a reign of terror.

By the time Pelsaert returned on September 17, 125 people had been slaughtered. Fortunately, Pelsaert was not taken by surprise. The soldiers on West Wallabi had managed to find water and food and were joined by fugitives from Beacon Island. They prepared for attack on the mutineers and succeeded in capturing Cornelisz himself. Pelsaert immediately brought the mutineers to trial. His detailed records of the trial testimony, known as the *Pelsaert Journal*, provides most of our knowledge of the atrocities on what became known as *Batavia*'s graveyard. Cornelisz and several others were executed on the spot; the survivors were taken on to Batavia.

Human remains from this time have turned up from time to time on Beacon Island since 1960. In 1994, the Western Australia Maritime Museum located a mass grave, and complete excavation of this gruesome find followed in 1999 and 2001. The museum has teamed up with forensic pathologists at the Queen Elizabeth II Medical Centre in Perth to investigate the nine skeletons found so far. The team's main focus is identification. Can the remains be matched with any of the known victims of the *Batavia* mutineers? To do this, the pathologists must determine as many details as possible about the victims – age, sex and physical characteristics, as well as cause of death – that can be compared with the historical records.

BELOW The mass grave excavated on Beacon Island. The remains of the six-year-old child, jammed between the back of the adult skeleton on the right and the edge of the pit, have already been removed.

At first sight, matching the remains with the 125 victims seems a daunting task. However, several factors make it easier. First, the population is quite diverse. The victims included men, women and children; they were from different countries and social backgrounds. Second, not all the victims were murdered on Beacon Island. Those killed on one of the other islands are unlikely to be buried on Beacon Island. Moreover, many were drowned and never buried. Pelsaert's *Journal* specifically notes whether bodies were buried or thrown into the sea. At least 10 people were buried on Beacon Island. The number is certainly higher, as the method of disposal is not always described. Third, the *Journal* often gives details that could be helpful for identification and is usually quite specific about how people were killed.

## The Human Remains

Four burials were found in the 1960s. The first was a young woman, 16 to 18 years old. A shallow indentation on the top of her skull showed she had been hit by a sharp implement. This would have stunned but probably not killed her. The skeleton of another young person, about the same age, was found in 1963. A musket ball in the chest area was presumably the cause of death. The skull had been misplaced since the original discovery, but another isolated skull is believed to be the missing one. Two men were found buried alongside one another in 1963 and 1964. One is a complete skeleton of a tall man in his mid to late thirties. He was hit on the head once with a sharp implement, probably a sword or cutlass. Only the skull of the second man has been excavated. He was in his mid-thirties and had been severely beaten about the head with a sharp instrument, possibly an ax or an adze.

The mass grave was a roughly circular pit, about 5 ¾ feet (1.75 meters) in diameter, containing the remains of three adults and three children, one of whom was an infant. The older of the children was placed in the pit first. It is difficult to determine the sex of juveniles, but the child was about 12 to 14 years old. Buttons were found in the chest region and close to the left wrist – the only evidence of clothing found. Two of the adults were interred next. The first was in fairly poor condition, and age and sex could not be established, other than that it was an adult. The second was poorly preserved and had been damaged and disturbed. Again, it was clearly adult, but age and sex could not be determined. The third adult was then placed on top. He was a man in his early twenties. He had bony growths on his left shin bone and, to a lesser extent, on his left foot. It is difficult to be sure of the cause of this, but it most likely resulted from some sort of infection, perhaps from an injury, and probably caused him considerable pain. His vertebral column shows evidence that he was involved in heavy physical work. He also had a

BELOW The right side of this man's skull was cracked, possibly from a blow to the head, and one of his teeth was forced into his jaw, also probably from a blow. He may have sustained these injuries at the time of death or in the shipwreck.

scoliosis, or lateral curvature of the spine. The body of the younger child, five or six years old, was interred last and pushed against the lower back of the third adult so that it would fit in.

Two skulls were also associated with the mass grave and must belong to the two adults whose sex and age could not be established. Unfortunately, the two skulls could not be matched with the rest of the skeletal remains so we do not know to which body each belongs. One skull probably belonged to an adult in his or her early forties. The second skull came from a man in his early thirties. He had a crack on the right side of his skull, probably from a blow to the head. His right upper incisor had been forced into the upper jawbone, most likely from a heavy blow.

A large mass of dark organic material was found in the central area of the burial pit. This was removed in 2001 and excavated in the museum's conservation laboratory. As well as several bones, it contained 19 small pre-erupted teeth from an infant about nine months old.

## Identifying the Victims

Is it possible to suggest names for any of the remains? The young girl discovered by chance in 1960 can be one of three people only. Few women were killed on Beacon Island, and three could have been in their teens.

One passenger, Maijken Cardoes, had a nursing infant. She was hit on the head in a struggle to escape; her throat was cut and she was buried in the hole dug for the family of the Predicant (the official minister appointed by the company). The age of the Predicant's middle daughter, Willemyntgie, is not known; she could have been in her teens or younger. She was beaten to death. The family's maid, Wijbrecht Claes, is described as "young" and could also have been in her late teens. She was stabbed to death. These three women were presumably all buried in a mass grave with other members of the family.

The young woman's burial seemed to be a single individual, but was not scientifically excavated. She was found only a few yards from the mass grave, and other remains may be in the area, perhaps in a second mass grave.

The person with the musket ball in his chest found in 1963 was probably Jan Dircxsz, one of the mutineers. He was shot in a fight with the soldiers on West Wallabi Island and died 11 days later, after Pelsaert had regained control. He was apparently the only person who died from a musket wound.

Clearly, the most likely candidates for the occupants of the mass grave are groups of people killed at the same time on Beacon Island. The *Journal* describes four such occasions.

BELOW The irregular surface on this *tibia* (shin bone) indicates this man suffered from a chronic inflammation below his left knee, resulting from an infection; it would certainly have caused him considerable pain.

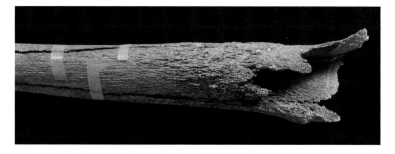

One was the murder of the Predicant's wife, six of her children, and the family's maid on July 21. Maijken Cardoes was also killed and buried with them. Hendrick Denijs, a company official, was also killed on the same night, but there is no record of what was done with his body.

A second group includes Jacop Hendricxsz, the carpenter; Paschier van den Enden, the gunner; and a sick cabin boy. All had their throats cut on July 10 or 12 and were buried. An English soldier, Jan Pinten, was also killed about the same time. Two further groups of sick people were killed on July 10 and 13.

The *Journal* describes no single group made up of three adults, two children, and an infant which matches the bodies in the mass grave. There are two possibilities. The people murdered on July 21 and the groups of sick people may be in more than one mass grave. Alternatively, other victims, murdered within a few days, could have been added to the grave dug for the two men and the sick boy.

Documentary information about the groups of sick people killed on July 10 and 13 is inadequate to match them with the occupants of the mass grave or any of the other burials. What information can be used to decide if either the Predicant's family or the group of two men and the sick boy are represented among the remains?

Despite the presence of two children and a baby who could be Maijken Cardoes's infant, the almost total lack of evidence for violence suggests that the mass grave does not contain the Predicant's family, who were all beaten to death with axes and adzes. Hendrick Denijs, killed the same night, was also severely beaten about the head. Jacop Hendricxsz, Paschier van den Enden, and the cabin boy were all stabbed and had their throats cut. These wounds would not necessarily leave traces on the bones. Moreover, Hendricxsz is described in the *Journal* as "limping." The skeleton of the man in his early twenties who suffered from chronic inflammation of the knee thus may be Hendricxsz.

Who could the other occupants of the mass grave have been? One likely candidate for the third adult is Jan Pinten, the English soldier whose throat was cut, probably on the same day as Hendricxsz, van den Enden and the cabin boy. A six-year-old girl, Hilletgien Hardens, strangled a couple of days earlier, could be the second child.

On the evidence available so far, neither possibility unequivocally fits the evidence from the mass grave. The research team is still exploring other lines of enquiry. DNA analysis would be informative. As well as allowing the two isolated skulls to be matched with the correct bodies, it would indicate sex and would show whether any of the occupants of the mass grave were related. If none of them were related, then they cannot be members of the Predicant's family. Unfortunately, attempts to extract DNA from the Beacon Island skeletal remains have so far proved unsuccessful. The team is exploring lead isotope analysis, another experimental technique. This can distinguish people from different regions and so could help identify the English soldier, Jan Pinten.

# High-mountain Inca Sacrifices

**M**ountain peaks are sacred, even today, to traditional Andean people. Their snow-capped peaks are important sources of water, and their role as *huacas*, or sacred shrines, dates back hundreds or perhaps thousands of years to the Inca and their ancestors. Today, as in the past, Andean pilgrims make offerings to high mountain shrines or gods. Half a millennium ago, when the Inca empire held sway, such offerings included not just small, valuable objects and animals but also, on especially important occasions, the most valuable of all things: beautiful and unblemished children and adolescent girls. Until relatively recently, such practices, while mentioned in historical accounts of the Inca, were not well documented archaeologically. There is now increasing evidence, however, that human sacrifice on the high Andean peaks was a widespread, if uncommon event.

The highest peaks in the Andes are a challenge to even the most experienced mountaineers. Spending time in these extreme altitudes is dangerous. This makes it even more remarkable that such settings were chosen for elaborate rituals by the Inca, but recent finds indicated clearly that they were. The most recent expedition to such a setting, the 1999 trek to the Llullaillaco volcano led by American explorer Johan Reinhard and Argentine archaeologist Constanza Ceruti, was a 23-day adventure. The team spent 13 days on the summit of the volcano at an altitude that

**BELOW** The high peaks of the Andes pose a tremendous challenge to climbers today, even with oxygen and climbing equipment. The Inca priests who led the Llullaillaco children to their sacred peaks had no such equipment, yet they regularly made the trek.

ABOVE Unbelievably, this is the face of a child who has been dead for more than 500 years.

BELOW A discovery of any kind during an archaeological excavation is tremendously exciting. When the first glimpse includes well-preserved textiles in what seems to be a tomb, excavators begin to hope for equally well-preserved human remains.

makes any work, even thinking, nearly impossible. Nevertheless, the team recovered some of the best-preserved mummies ever found in the Andes and added significantly to the catalog of known frozen mummies of Inca sacrifice victims.

The Llullaillaco volcano, in northwest Argentina, is located in what was once a relatively remote province of the Inca empire. Yet the rituals performed there were clearly central to the empire, and they must have been sponsored in some way by the central government in Cuzco, Peru. This became clear to the expedition team when they reached the summit of the volcano, 22,057 feet (6,723 meters) above sea level. There, in the bitter cold and constant wind, they found evidence of extraordinary events that occurred more than 500 years ago.

At the summit, the Inca had constructed two stone shelters and a platform surrounded by a stone wall. The platform, measuring about 33 by 20 feet (10 by 6 meters), turned out to be the central site for the Llullaillaco sacrifices. Inca offerings, in the form of small figurines of gold, silver and precious *Spondylus* shell (a material that comes from coastal Ecuador), signaled its importance. Excavations under the platform revealed three separate tombs, each containing the frozen remains of a sacrificial victim. The mummies comprised a little boy, a little girl and an adolescent girl. They were in near perfect condition, except for the little girl, whose corpse had been struck by lightning. Even this detail was significant; it may indicate that part of the ritual conducted on the volcano was dedicated to a lightning god.

Studies of the mummies began almost immediately. Soon after being brought down from Llullaillaco, and without being unwrapped or undressed, the mummies were subjected to CT scans (computed tomography imaging, also known as CAT scanning or computed axial tomography), which shows not only bones but also soft tissue and blood vessels. Remarkably, the scans appeared to demonstrate that the bodies of

the Llullaillaco mummies still contained frozen blood. This made them unlike the other 15 mummies that have been found on high Andean peaks, all of which were more freeze-dried than frozen, and none of which contained frozen blood. The Llullaillaco bodies showed none of the evidence of cell damage seen in the rest of the Andean mummies. Their outstanding state of preservation made these some of the most scientifically valuable mummies yet found.

Their excellent condition was confirmed by radiographic analysis. The X-rays of teeth and bones allowed researchers to estimate the age of each mummy at death. The little girl who had been struck by lightning was around 6 years old, the boy around 7 years old and the older girl around 15. The internal organs of all the mummies were intact. The lungs were still filled with air, and blood was visible in the hearts and blood vessels. None of the child mummies showed any sign of trauma or injury of any kind, eliminating the idea that they were killed with a blow to the head, as was known to have happened to other Inca sacrifice victims. Another finding of interest was that the head of the older girl had been deliberately deformed, as was common among elite members of various ethnic groups. To make the head conical, as this girl's head was, an infant's head is wrapped so that the skull takes the form of a cone (or a mountain peak).

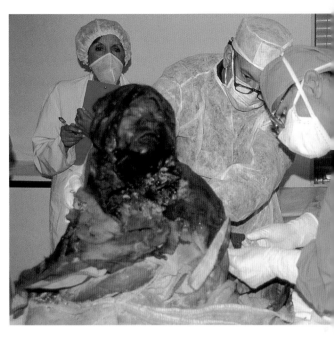

The next study conducted on the mummies was analysis of the hair, which often holds residues of narcotics ingested by people in the weeks before they died (although anything taken in the several days just before death is unlikely to show up in the hair). The adolescent girl showed signs of having taken a great deal of coca, the plant that is the main raw material for cocaine; in its unprocessed form, coca is an important ritual and medicinal plant in the Andes.

An important question concerning the victims of sacrifice is who they were. DNA analysis of samples from the Llullaillaco mummies is providing a first step toward answering that question. Analysis of their mitochondrial DNA indicates that they are clearly related to groups of ancient and modern people of South America. Until more comparative

**ABOVE** The practically perfect preservation of the frozen mummies of Llullaillaco allowed a wide range of sophisticated analysis to be done, which is why we know so much about the children who were sacrificed in the more remote regions of the Inca Empire.

**LEFT** These mummies are so well preserved that they almost appear to be alive. Such incredible preservation allows research into a wide range of topics, from ancient hairstyles to stomach parasites.

121

ABOVE The child mummies at Llullaillaco are so well preserved that even experienced archaeologists are startled to see faces and hands that look almost alive.

DNA data are available for specific groups, however, it is difficult to pinpoint the known prehistoric ethnic groups from which the mummies came. Analysis of nuclear DNA did suggest that the two girls from Llullaillaco may have been closely related, although they did not have the same mother.

The scientific analyses of the frozen remains of the Llullaillaco mummies, together with the archaeological information about the site they came from and the items found with them, can be combined with the historical information we have about Inca sacrificial practices in an attempt to understand what might have happened to these children.

The accounts we have of Inca beliefs indicate that human sacrifice was a rare and extremely sacred ritual, one reserved for the most somber of occasions. Those chosen for sacrifice were the most beautiful and unblemished children and the most beautiful and pure young maidens. It is probable that having a child chosen for sacrifice was a great honor for a

RIGHT Gold and silver figurines of both males and females dressed in the Inca style with feather headdresses have been found in several of the high-mountain sacrifice sites of the Andes.

family or a village. Those chosen from the provinces of the empire (or at least some of them) were taken on a pilgrimage to Cuzco, where they participated in specified rituals and were anointed in some way. They were allowed to don clothing and adornments normally reserved for the Inca elite, and they traveled in a style that was likewise limited to the elite. They were taken to the summit of the high mountain in a procession, led by Inca priests, that featured feasting, music and multiple rituals along the way. There, in the culminating ritual, they were killed or allowed to die. The sacrificial victims were plied with coca and perhaps narcotics, probably for hours and possibly days before they died.

**ABOVE** The excavation of these mummies was complicated by bitter cold, little oxygen, and almost constant storms.

Drawing on this historical information, it appears that the three victims from Llullaillaco were sacrificed in an elaborate ceremony under Inca state sponsorship, which was led by religious leaders of the empire. The archaeological information and the items found with the mummies confirm this idea. The young sacrificial victims were all dressed in fine clothing made of alpaca wool textiles, some in bright colors (red and yellow). The adolescent girl wore a white feather headdress. A huge range of precious offerings was found both inside the tombs of the mummies and in and around the sacrificial platform above the tombs. These included a necklace made of *Spondylus* shell and fancy Inca textiles, including those found wrapping or being worn by the mummies and in two separate bundles. Thirty-six carved figurines made of gold, silver and *Spondylus* shell were found at the site. These included a large set of llama figurines aligned to depict a caravan and a series of dressed human figurines, some with inlay. There were elaborate decorated Inca pots, some containing food offerings, and many with intact lids. The range and number of offerings suggest an important and perhaps a large group had a hand in providing the offerings.

The Llullaillaco mummies are the best preserved of the frozen mummies that have been found in the Andes. Continuing study will reveal information on their health, nutrition, genetic relationships and perhaps many other aspects of their life. Their significance, however, lies not only in their specific study but also in the possibility of comparative studies with other mummified victims of Inca sacrifice. The Llullaillaco mummies, along with the famous remains of the mummy nicknamed Juanita and others from Peru and Chile, bring to a total of 18 the scientifically known Inca sacrificed mummies from the Andean region. The ongoing study of these remains continues to open frontiers in the contribution such bodies can make to our understanding of the past.

# Prehistoric Homicide and Assault

**W**e think of assault and violent death as signs of the times in which we live. However, a series of skeletons a couple of thousand years old from the coast of the Western Cape Province of South Africa reminds us that people have been attacking and killing each other since the Stone Age.

## Melkbosstrand

### A Gruesome Discovery

In July 1996, a construction worker excavating a sand dune with a backhoe in Cape Town's northern coastal suburb of Melkbosstrand was astonished when human bones appeared in the wall of his trench. The police were summoned, but they turned the case over to archaeologists John Parkington and Royden Yates of the University of Cape Town when they realized the remains might be of precolonial age.

The suspicions of the police were confirmed when the archaeologists found a double burial covered by a large stone, possibly a capping stone or grave marker, 4 feet, 3 inches (1.3 meters) below the modern surface and below an undisturbed precolonial rubbish dump of shellfish remains. There were no grave goods. Features of the skulls were typical of the indigenous hunter-gatherer people of the region, and radiocarbon dating of the bones showed that both skeletons were about 2,500 years old. The skeletons were lying side by side on their backs with their legs bent at the hips and knees; their arms were bent at the elbows with the hands toward

BELOW Two Stone Age homicide victims with fatal head wounds from Melkbosstrand, Cape Town, South Africa: the skull on the left belonged to a teenager, while that on the right was a woman in her twenties or thirties. They were buried together.

the face. Unlike the rest of the bones, both skulls were stained black. This is probably a consequence of chemical changes in ocher that occurred after death. Ocher is a soft iron oxide pigment widely used as decoration, which they might have had in their hair. Both skulls had clearly been damaged when fresh.

Because different kinds of foods have characteristic carbon and nitrogen compositions, the carbon and nitrogen content of bones and teeth can indicate aspects of diet. Such studies of the Melkbosstrand individuals suggested that they ate either a great deal of coastal marine shellfish or the kinds of plants found in the interior, as well as animals who ate these kinds of plants. A shellfish diet seems more likely, because people who lived along the Western Cape coast in the millennium between about 3,000 and 2,000 years ago collected shellfish and created huge dumps of their remains, called megamiddens.

## The Remains

One of the skeletons, Melkbosstrand 1, is complete. It belonged to a teenager; the teeth and long bone development are typical of a 13- to 16-year-old. Although still immature, the width of the greater sciatic notch, an indentation on the lower border of the hip bone, strongly suggests the skeleton is of a female. Her thigh bone length of 15 inches (37.9 centimeters) correlates with a height of about 4 feet, 8 inches (142 centimeters). Her teeth are in good condition, with no evidence of cavities, and there is no sign of disease or breakage on her long bones, but the presence of slight porous bone growth in one of her eye sockets, known as *cribra orbitalia*, indicates she had a degree of iron deficiency.

Her skull has two circular breaks with radiating cracks that must have occurred when the bone was fresh, because the way the bone is bent inward is typical of what happens when fresh bone suffers high-impact damage, and because there is no sign of healing. The gashes were made by blows rather than cuts, probably by hitting with an implement with an uneven edge. Magnification of one of them shows a scratch mark and depressions consistent with damage from a stone tool.

The second skeleton, Melkbosstrand 2, is complete except for the end of one of the arm bones, which was probably lost during discovery. The hip bones clearly indicate that it belonged to a female. The degree of pitting and scarring in telltale places on her hip bones is more than one would expect from age alone and probably reflects the trauma of giving birth to one or more children. Her thigh bone lengths of 15½ and 16 inches (39.5 and 39.9 centimeters) suggest she was about 4 feet, 10 inches (148 centimeters) tall. The wear on her teeth is consistent with that of a hunter-gatherer in her twenties or early thirties, while the degree of fusion of her bones reflects an age between 24 and 46 years.

**ABOVE** A piece of bone bending inward on the lower border of the blow on the left indicates that the damage to the skull of the Melkbosstrand teenager occurred while the bone was fresh. The line just above the scale is a long gash shown in close-up below.

**BELOW** A close-up view of the gash on the back of the skull of the Melkbosstrand teenager shows a scratch mark parallel to the upper edge and two depressions within it which suggest that it was made by a stone tool. The scale is in millimeters.

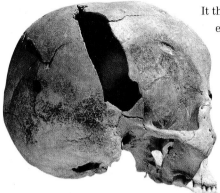

ABOVE The right side of the skull of the female from Melkbosstrand shows a depression toward the front, two large slashes, 3.1 inches and 1.2 inches (80 and 30 millimeters) long respectively, further back, and a circular wound below the second slash.

It therefore seems likely she was in her thirties. Like the teenager, her eye sockets indicate mild iron deficiency. Her lower back bones show at least one herniated and one slipped disc, which would have caused lower back pain and limited her flexibility. She also seems to have suffered trauma at birth that restricted blood flow and left her with an asymmetrically developed neck, upper body and left arm. Her left shoulder blade is about a third smaller than the right one. This did not necessarily cause her severe disability, as there is no evidence of disuse.

Her skull was marked by cracks extending from two large slashes to the head and a small puncture wound probably made by a pointed projectile hitting fresh bone. There is also evidence of damage above her right eye and two other depressions on her skull, one on her right temple and one on the left side made by forceful blows on the opposite side. She was in all likelihood knocked unconscious by the blow to her right temple, then hit on the right side of her head while the left side was resting on a surface.

### A Verdict of Homicide

It seems that the teenage and adult females were deliberately put to death, as there is nothing in the burial site that could otherwise have caused the damage to their skulls. It has been suggested that they were killed when asleep, unconscious, or tied up, because there is no evidence of wounds from defensive actions, such as might be found on their arms. Unfortunately, the archaeological record has left no clues about the circumstances that led to this violent event.

## Snuifklip

### Rescued Remains

Another skeleton of roughly the same age from the Western Cape Province coast also has evidence of deliberately inflicted head wounds, but the victim survived them. The remains were excavated in 1984 at Snuifklip, a well-known fishing spot near Vleesbaai on the southern coast, by Francis and Anne Thackeray, then of the University of Stellenbosch near Cape Town, after the owner of a holiday home reported a skull eroding out of a sand dune on his property and expressed concern that it might be removed by passersby. The bones were described by Alan Morris of the Department of Human Biology at the University of Cape Town.

### The Grave

The skeleton was found in a crouched position with the right arm between the legs. The bones are extremely brittle as a result of burial in loose sand. Almost all the bones were recovered, except for a few finger bones. A pile of grave goods was found in front of the face, including a hippopotamus tooth and a fossilized back bone of a large grazing animal;

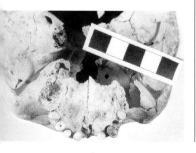

BELOW This close-up view of the front teeth of a 2,400-year-old elderly male from Snuifklip on the southern South African coast shows that they are almost completely worn down. The scale is in centimeters.

limpet and mussel shells, one with a hole through it, possibly for stringing as a pendant; and stone tools. Radiocarbon dating indicated that death occurred some 2,400 years ago.

## The Skeleton

The remains are characteristic of the indigenous hunter-gatherer people of the region. Considerably worn teeth, the state of fusion of the bones, and evidence of degenerative joint disease indicate an adult individual of advanced years, probably older than 50. The identification of the sex is difficult because the hip bone is poorly preserved, but the development of the brow ridges and margins of the eye sockets as well as the existence of well-defined muscle markings on the bones strongly suggest the person was male. Interestingly, he was of short stature, as his long bones suggest he was only about 4 feet, 10 inches (148 centimeters) tall.

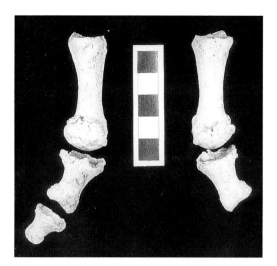

ABOVE The bones of the big toes of the Snuifklip individual show how they have been bent outward because of arthritis. The scale is in centimeters.

Evidence for degenerative joint disease is particularly noticeable in the bones of his lower back and feet. Fused vertebrae in his lower back must have caused considerable pain during the fusion process, although this would not have seriously inhibited his mobility. However, the arthritis in his big toe was so severe he must have had great pain and difficulty walking any distance.

Not only are his teeth very worn but also there is evidence of cavities, abscesses and loss of teeth. Some of these were lost just before death but others, whose sockets had disappeared and were completely resorbed into the jaw bone, were lost years before death. The state of his teeth must have caused him great discomfort and is surprising for a Stone Age hunter-gatherer, as this way of life is usually associated with a low incidence of cavities. However, Morris has observed high rates of cavities in other prehistoric burials from the area and suggests it may be a consequence of the lack of fluorine that has been documented in the local water.

Like the Melkbosstrand skeletons, *cribra orbitalia* in the eye sockets indicates that the Snuifklip individual also suffered anemia. Generally, he seems to have been in poor health for some time before he died.

## Head Wounds

The Snuifklip skeleton has two head wounds on top of the skull. One is a small, roughly circular area of damage that crushed the diploë, the area of spongy tissue between the outer and inner plate of skull bone, but caused almost no depression into the interior of the skull. The second area of damage is larger and more elongated in shape, also crushed the diploë, and additionally pushed the inner plate of skull bone in toward the brain, although no sharp bone fragments intruded. Because the wounds

BELOW The top of the skull of the Snuifklip individual shows two depressions, indicating that he suffered head wounds that healed before he died.

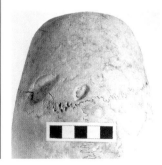

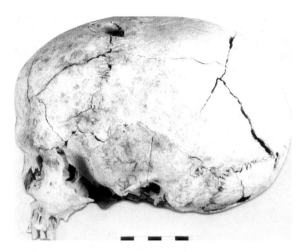

**ABOVE** Left-side view of the Snuifklip skull, which is slightly distorted because of damage caused by cracks after being buried in loose sand.

show signs of healing, the victim evidently survived the trauma. They in fact have a state of healing similar to that of a known individual who suffered a nonfatal head injury about six months before he died. Both wounds on the Snuifklip male are very localized and were probably made with a small object that connected with the skull with great force. It is therefore unlikely that they were accidental.

## Quoin Point

### A Chance Find

One day during the summer of 1966–1967, archaeologist John Parkington of the University of Cape Town and Geoffrey Voigt were searching a sand dune area near Quoin Point, on the southern coast of South Africa, southeast of Cape Town, when they noticed a scatter of bones that had eroded out of a dune. Careful sieving of the sand led to the recovery of many of the bones of an adult and a baby – also some animal bones, mostly of dune mole rat, and three thin-walled potsherds with coarse temper typical of the kind made in the region during the past 2,000 years. The bones were studied by Alan Morris of the Department of Human Biology at the University of Cape Town.

### The Skeletons

The hip bone of the adult skeleton is undoubtedly that of a female, which is confirmed by her light build and absence of massive muscle attachments. The state of fusion of her bones suggests an age in her late twenties, which is supported by an age in the range of 22 to 40 years indicated by the condition of her pubic symphysis, the area where the hip bones meet in front. Scarring on her hip bones suggests she may have had one or two children.

Her long bones indicate that she was short, between about 4 feet, 9 inches and 5 feet (145 and 152 centimeters) tall. The skeleton of the child is less well preserved but suggests an age of less than three months. Even if this is a slight underestimate, given the smallness of the indigenous hunter-gatherers of the region, the child was so young that it would still have been completely dependent on its mother's milk.

### Death by Bone Point

One of the adult vertebrae (bones of the back) has fragments of two bone arrow points actually embedded in it. The smaller of the two embedded pieces is broken at both ends. It is a short section of the shaft of a bone point that is cylindrical in cross section and typical of those found in archaeological sites in the region. Two more such pieces were recovered by the archaeologists through sieving; one has a flat base characteristic of

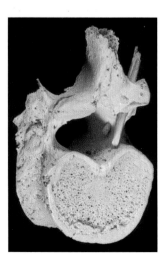

**BELOW** An oblique view of the broken pieces of bone points embedded in the back bones and entering the vertebral canal of a Stone Age woman from Quoin Point, southern South African coast.

the rear end of bone arrow points. Although they do not fit directly onto the fragment in the vertebra, these pieces are likely to have been part of the same point. The larger embedded fragment is flat in cross section and could have been part of the kind of bone point with a tang or projection at the base for attaching it to a handle, as is known from historical examples in museum collections.

The fragments probably represent two separate missiles. The severity of the resulting punctures suggests they were fired at close range, and that there were two of them is probably an indication that the wounds were not accidental. The trajectory of the angle of entry of the larger fragment suggests that the attacker was either kneeling or crouching behind a standing victim, or the victim was lying face down with her attacker standing over her. The angle of entry of the smaller fragment is slightly different; and might have been deflected by a rib.

The arrows of traditional hunter-gatherers like Bushmen are not intended to kill as a result of trauma but are rather delivery mechanisms for introducing fatal poison made from the larvae of certain kinds of beetles. The Quoin Point female's wounds caused by the bone points were not necessarily fatal in themselves, but if the points were poisoned, death would have been inevitable. Certainly, there is no sign that the damaged bone had time to heal before death.

Interestingly, the young child's skull shows signs of bone resorption suggestive of acute starvation. If the bone point victim was the child's mother, she would probably have been unable to feed her infant even if she lingered for days before the poison finally killed her.

**ABOVE** Side view of the bone projectiles embedded in the back bones of the woman from Quoin Point.

**BELOW LEFT** This is how two bone projectiles entered the back bones of the Quoin Point woman. One deflected off a rib.

**BELOW RIGHT** Line drawing of the profile of the Quoin Point woman, showing the position and estimated route of entry of the bone points.

## Violent Times

The Melkbosstrand, Snuifklip and Quoin Point cases record rarely glimpsed events of interpersonal violence in the Stone Age archaeological record. Unfortunately, we cannot reconstruct the scenarios that led to the dramatic moments preserved by the bones, nor do we even know whether the assailants and victims belonged to the same group. Nevertheless, these examples serve to remind us that people in Stone Age times also experienced difficulties with conflict resolution and that assault and homicide are not confined to complex, techno-logically advanced societies.

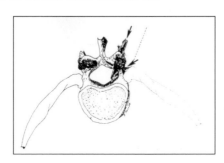

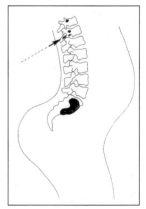

# The Sacrifices at
## Huaca de la Luna

I t is rare that a discovery sends shivers down the spines of archaeologists, but that is the effect that some recently uncovered human skeletal remains have been having. The remains include whole skeletons, partial skeletons, and the bones of dismembered arms, legs, hands, and feet from more than 70 young men who appear to have been ritually sacrificed some 1,300 to 1,500 years ago. They were found on a platform at the Moche ceremonial site of Huaca de la Luna, on the north coast of Peru, and researchers are still attempting to fully understand what happened there and why.

Huaca de la Luna is one of two major pyramids at the Moche site, the type site for the Moche culture of north coastal Peru, which thrived between roughly AD 200 and 700. The site is known for its monumental adobe brick architecture decorated with fabulous painted walls, which tourists visit in throngs each year. Research has also revealed extensive ancient residential compounds where Moche people lived and worked. It is only recently, however, that researchers found the sacrificial platforms where grisly rituals, including the sacrifice of large numbers of captured war prisoners, were conducted. Some evidence that cannibalism may have been practiced in some cases has been identified.

The most recent research effort at Huaca de la Luna took an unexpected direction when archaeologists encountered human bones scattered across a plaza near the base of a natural stone outcrop. As they cleared the area, the archaeologists found that the bones covered an area

**BELOW** The articulated (connected) skeletons and partially disarticulated individuals found by archaeologists at Huaca de la Luna were studied in the laboratory and found to belong to young men who had suffered multiple injuries in combat.

of roughly 66 square yards (60 square meters); as they cleared them off, they found that few of the bones were in their natural, or articulated, state. Rather, the skeletons of most of the individuals had been dismembered. Most of the skeletons were missing their skulls or one or more of their limbs or extremities.

During continuing research, excavators found a series of hard clay surfaces, some with bones on top of them, others with bones embedded in the clay. They also found an ancient pit, about 4 yards (3.6 meters) square, dug into a plaza and a wall, that was filled with a combination of human bones and broken pottery effigy vessels in the form of nude human males with ropes around their necks. All of the human bones were associated with clay surfaces that appeared to be the result of deposits of construction clay that was washed into the plaza during episodes of torrential rain.

Understanding this discovery required further study of the bones as well as pulling together other lines of evidence as to what the Moche were doing at Huaca de la Luna. Laboratory analysis of the human remains was challenging because there were so many skeletal portions and so few complete skeletons. This made it difficult to determine precisely how many individuals were represented. Careful study revealed that at least 70 people were represented by the bones, although the number may have been somewhat higher. Analysis of age and sex, usually a basic step in the analysis of human remains from archaeological sites, revealed the first clue as to who these people were. All of the bones belonged to males between 15 and 39 years of age; the average age at the time of death was 23.

Examination of physical indicators of health showed that the men and boys were universally healthy and physically active, and that

**ABOVE** The way the limb bones were spread and not always in anatomically correct position suggested that the bodies of the sacrificed warriors may have been left exposed for the vultures.

**BELOW** The more complete skeletons provided the most solid information on the age, sex, and health of those buried at Huaca de la Luna. There were many individual bones that were not connected to full skeletons, making it difficult even to be sure how many individuals had been sacrificed.

they had enjoyed excellent nutrition from childhood through to the time of death. Their bones did show that not all of their activity was benign, however. Many of the bones showed evidence of healed fractures, including fractures of arms, legs and skulls. The pattern of the healed breaks was relatively consistent and appeared to indicate that they were almost all the result of interpersonal violence, with a specific set of weapons, and were not in any sense accidental. Many of the bones also showed evidence of relatively new injuries that had apparently occurred within about a month of the time the individuals were killed. These included broken limb bones just beginning to heal as well as injuries to the nasal area.

The analysis of sex, age, general health and past physical trauma indicates that this was a special group of young men. There have been indications for some time that there was a specialized group of warriors under the Moche. The discovery of these sacrificed males certainly supports that idea.

## Moche culture

Evidence for how the men and boys died was seen in bone damage that showed no signs of healing. Few, if any, appear to have actually died in combat. Instead, many of the skeletons showed signs of cut marks on the neck bones that are consistent with cutting of the throat, while several others had massive skull fractures, as though they had been struck with blunt objects such as war maces.

**BELOW** Many of the bones were deliberately cut, like the bones shown in this picture. Also, like this skull, many were found without the mandible (lower jaw).

As mentioned above, few of the skeletons were intact. Most of the skeletons were missing limbs, extremities or skulls. Also, many severed limbs were found, not all of which could be clearly linked to a specific skeleton. The majority of the remains, however, showed no evidence of how the bodies had been dismembered. The bones showed no cut marks or signs of forceful separation while soft tissue was present. Many explanations are possible. One is that the dismemberment of the bodies happened in a haphazard manner, after death, due largely to scavengers such as vultures. Many of the bones were at least partially sun-bleached, indicating that they had been exposed to the elements for some time, rather than buried or

protected. It is also possible that the bones were deliberately scattered in the pattern in which they were found after the exposed bodies had deteriorated.

One set of bones, however, did show evidence of dismemberment. A group of skeletal remains of six or seven young men was found in a plaza adjacent to the one where most of the human remains were located. This group of bones, in addition to showing many of the same characteristics as the rest of the bones, showed evidence of cut marks consistent with cutting through muscle not only to dismember but also to deflesh the bone. The pattern of the cut marks is not very different from that seen on the bones of animals butchered for eating. This suggests the possibility that the people in this group were subject to ritual cannibalism.

The evidence from the bones alone suggests massive episodes of human sacrifice by the Moche, most probably of young male warriors who had been taken prisoner. When this evidence is added to other lines of evidence, we can develop a much richer idea of what happened long ago at Huaca de la Luna. Both Moche art and information about the environment and the climate provide clues as to what occurred.

There is widespread evidence in Moche art and iconography that warfare and the holding and execution or sacrifice of prisoners were rituals of major importance to the Moche. Wall frescoes at Moche sites show processions of naked male prisoners with ropes around their necks. Pottery vessels made in the form of effigies of apparent prisoners are common; they likewise depict naked males with ropes around their necks, sometimes with hands free, sometimes with hands tied. The fineline painted pottery of the Moche, however, holds the most detailed clues about the lives and deaths of the sacrificial victims at Huaca de la Luna.

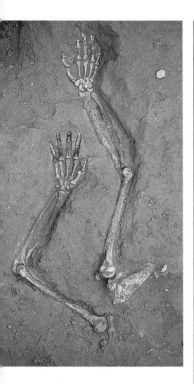

ABOVE Disarticulated (disconnected) bones of limbs, like these two sets of arm bones, were found in several areas.

Scenes from Moche fineline painted pottery include everything from hand-to-hand combat using maces to warriors dressed in ritual attire to races and hunts conducted by young men, possibly warriors. They also include elaborate scenes that researchers believe depict the arraignment and sacrifice of prisoners. Such scenes show naked men being presented to a highly adorned individual sitting in a position of honor, and they often feature a lower register showing naked men being killed or having been sacrificed. Vultures, or men dressed as vultures, serve as guards and executioners. Other scenes depict the victims of sacrifice, decapitation and dismemberment, and dismembered limbs with rope around them. To the Moche, prisoner sacrifice was clearly a significant ritual theme.

In another thread of evidence, rain, considered beneficial in many areas of the world, was probably a catastrophe to the Moche. Their environment, the north coast of Peru, is an extremely arid region with almost no measurable rainfall during most years. Agriculture, which along with fishing was a mainstay of the Moche economy, was focused in the rich river valleys that cut across this desert region. The Moche, like all the ancient civilizations of Peru, invested huge amounts of labor and resources in building major networks of irrigation canals. Rain in any quantity occurred only occasionally, the result of the periodic climatic disturbance known as El Niño. Niño events cause catastrophic rains that destroy canals and agricultural fields even as they disrupt the normal ocean patterns and cause major dips in fishing productivity. Thus, the various layers of rain-washed clay seen in the area where the sacrifice victims were found suggest that the sacrifices took place during what were, for the Moche, catastrophic times.

Shortly after the sacrificial area was constructed, and during a period of rains, five or six individuals were sacrificed, and their bodies or skeletons were deposited in a pool of mud and left exposed as the mud dried. Just after another episode of rain, a much larger group of prisoners was sacrificed. Then, a pit measuring 4 square yards (3.3 square meters) was dug and the jumbled remains of multiple individuals, as well as many smashed clay vessels in the form of prisoners, were buried in it. Some time later, and after more torrential rains, many more individuals were killed, decapitated and dismembered. The corpses were left, in many cases one on top of another, exposed to the elements – including, most probably, vultures.

The significance of such a set of sacrificial rituals to the Moche must have been profound. That they took place during an episode, or set of episodes, of torrential rain suggests that a large El Niño event, or even back-to-back El Niño events, may have occurred. In such a case, the Moche may have been in serious trouble, and hunger, disease and warfare may have been rife. The prisoner sacrifices perhaps were meant to appease the gods and to bring back the normal climate. It is not clear, however, that they did so, as little additional construction at Huaca de la Luna occurred after the sacrifices took place.

# CHAPTER FOUR

# burials

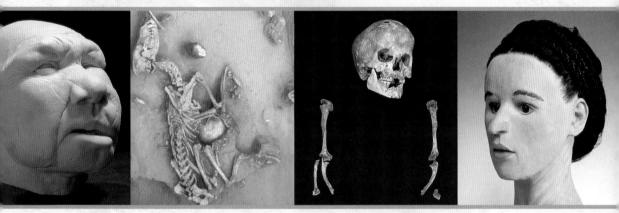

T he vast majority of human remains from the past take the form of purposeful
burials. While most of these are run-of-the-mill cases of little individual interest,
some are truly extraordinary for different reasons. Most books on tombs focus on
glamorous grave goods and treasures. Here, the focus is on some fascinating and
sometimes even startling cases where the bodies themselves speak volumes.

# Pit of the Bones

he Sierra de Atapuerca, located 9 miles (14.5 kilometers) east of Burgos in northern Spain, contains a veritable treasure-house of prehistoric sites, which have provided some of the earliest evidence for human occupation of Europe; the Gran Dolina (Big Sinkhole) site, for example, is a collapsed limestone cave that has yielded stone tools dating back a million years, as well as over 100 fragments of several individuals belonging to a kind of primitive fossil human named *Homo antecessor*, which lived here between 800,000 and 1,000,000 years ago.

Even more extraordinary is the Sima de los Huesos (Pit of the Bones), which has proved the world's largest known repository of fossil humans from the Middle Pleistocene period, between 780,000 and 127,000 years ago. In fact, it contains more human remains from this period than all other known sites combined.

To reach the pit is an adventure in itself. The sierra contains an enormous cave, the Cueva Mayor, and a journey into it involves clambering over rocks and up and down clay slopes, and crab-walking, going on all fours, or crawling through small spaces between stalagmites.

**BELOW** A view of the excavations in the upper levels of the Gran Dolina, one of the oldest archaeological sites in Europe, and one of the most important in the world.

After about 547 yards (500 meters), one reaches a chamber about 39 feet (12 meters) high; a side chamber contains cave bear nests and claw marks gouged in pockets of clay, which look as fresh as if they had been done yesterday.

There is also a great deal of graffiti, the oldest dating to AD 1561. Local people have been coming into the cave for centuries to obtain bear bones and teeth. According to tradition, every young man of the region must present his fiancée with a bear tooth from the Cueva Mayor to prove his prowess.

## Ongoing Excavations

In 1976, Trinidad Torres, a Spanish paleontologist, came here to collect sediment containing bear bones, but in it he also found an archaic human jaw, which triggered a campaign of excavations that continues today. In 1983, a test excavation was carried out at the jaw site by Juan Luis Arsuaga, who then began work in earnest the following year. During the first six seasons, his team removed countless blocks fallen from the cave roof and eight tons of disturbed sediments that had been trampled over the centuries, thus exposing the top of the pit.

ABOVE It took about fourteen days of painstaking work to remove the clay from around skull number 4 and extract it from the earth which had enclosed it for so long. The work had to be done incredibly slowly and carefully because of the extreme fragility of the remains.

By 1990, excavation began in intact deposits at the pit. At the top are thousands of bear bones. These decrease in number as one descends, becoming mixed with human remains, which predominate at the base and which date to more than 200,000 years ago. This means that presumably, sometime after the pit was no longer used by humans, hibernating bears began to fall into it. Doubtless their blundering about accounts for much of the disarticulation and fragmentation of the human remains.

Descending into the pit today involves grappling with a rope ladder nearly 39 feet (12 meters) long. At the bottom, after a further descent of several yards along a clay slope, one enters a small chamber about 13 by 7 feet (4 by 2 meters). Here, every summer, five or six people work. Some carefully brush the clay off bones while others draw and take measurements. Until 1995, the diggers had to work stooped or crouched, but since then it has been possible to stand upright.

The work is extremely painstaking. Each year, in one month of excavation, the team digs down only 10 inches (25 centimeters) and usually recovers about 300 human bones. The specimens are soft and fragile while still wet but, once dried and treated with preservative, they become quite hard. In each square, the smaller and lighter bones are found at the top and the heavier ones below. The laboratory in Madrid now contains drawers filled with tiny phalanges and other hand and foot bones, hitherto virtually unknown from this period. More remarkable is the fact that, when sediments excavated from the pit are sieved in a nearby river, the team can even recover the tiny bones of the inner ear.

**RIGHT** Frontal and maxilla of a child of *Homo antecessor*, about 12 years old, found in the Gran Dolina.

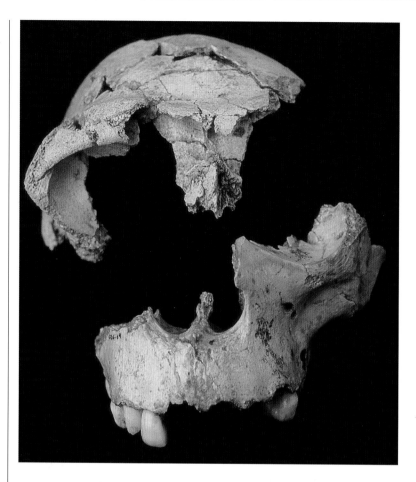

**BELOW** The first reconstruction ever made of *Homo antecessor*, this represents the child from the Gran Dolina

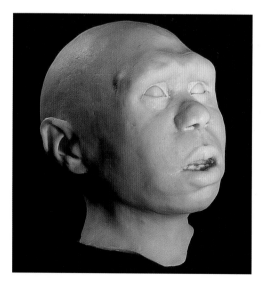

The Pit of the Bones has revolutionized our knowledge of archaic *Homo sapiens*, a transitional species between *Homo erectus* and Neanderthals. Although only a few square yards – maybe 4 percent – of the pit's deposits have been examined, more than 2,000 human bones have been found. Studies of teeth indicate that the human bones here come from at least 33 individuals and possibly 50. The bones are mixed up, but all parts of the skeleton are present, and males and females are equally represented. Most are adolescents and young adults aged between 13 and 22, and more than 30 percent are between 17 and 19; the youngest is about four and the oldest 35.

## The Significance of the Bones

The bones represent about 90 percent of all pre-Neanderthal remains ever found in Europe. These people seem to have been robust and quite tall. Their teeth are worn from chewing plants, but they had no cavities;

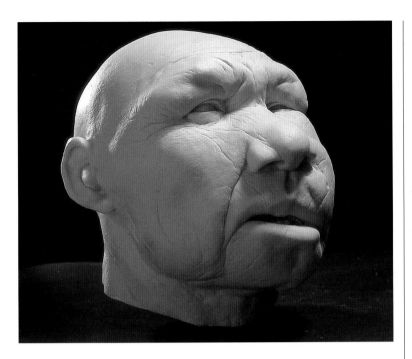

LEFT A reconstruction made from skull 5 from the Pit of the Bones, an adult *Homo heidelbergensis*. This remarkably complete skull has made possible our first reliable glimpse of a human face of 300,000 years ago.

BELOW The famous skull number 5 from the Pit of the Bones, found in 1992; the jaw was found the following year. It is the best preserved skull in the history of palaeontology, has a small cranial capacity (1,125 cubic centimeters), and is about 300,000 years old.

overall, the bones show few signs of illness or trauma. The remains include three remarkably well preserved skulls, found in 1992, with large Neanderthal-like brow ridges and projecting faces, though different from Neanderthals in overall shape. One skull has a capacity of 84.8 cubic inches (1,390 cubic centimeters), which is bigger than *Homo erectus* and early *Homo sapiens*, and lies within the range 73.2–103.7 cubic inches (1,200–1,700 cubic centimeters) of modern populations. The jaw matching this specimen has recently been fitted to it, and there is also a complete pelvis.

Since less than a quarter of these individuals lived beyond their early twenties, they cannot represent a full population, and it is probable that the older people were disposed of elsewhere. The absence of herbivore bones and stone tools indicates that this was not an occupation site, while the lack of carnivore damage suggests that the bones were not left there by predators. Arsuaga believes that over several generations, bodies were carried into the cave from an entrance, now lost, near the pit and thrown into the shaft in a form of mortuary ritual that may point to some embryonic religious belief – the oldest known funerary practice in the world.

# The Prehistoric Graves of
## Siberia

T he Baikal area of Siberia is a vast territory around Lake Baikal, with a varied relief including mountain ranges, lowlands and plateaus. They form the watersheds of such major rivers as the Angara, Lena, Selenga and others. The rich local flora and fauna, as well as the favorable climate, have drawn people to the area since ancient times. The famous paleolithic sites of Mal'ta and Buret are located here. In addition, the archaeological remains of the Holocene period, represented by multilayered site complexes, are also of great interest. The Baikal area is unique in that it is the only region of the vast territory of Northern Asia where hundreds of Final Mesolithic and Neolithic graves have been discovered and studied. Some of these are of enormous interest.

## Late Mesolithic Burials

Two chronological periods are known in the Baikal area. They are:
• the early period, or Khin period, of about 7,700 – 8,000 years ago,
• the late period, 7,200 – 7,400 years ago (uncalibrated radiocarbon age).

The early or Khin period includes burials with characteristic structures above them; the body is extended (in three cases) or flexed on its side (one case). They are oriented northward and sometimes contain ocher. The late period comprises 16 burial complexes, all of which have been opened and examined. They are characterized by stone structures above and inside the grave; the bodies are flexed on their side or are supine with their knees bent; they are intensively painted with ocher. The grave goods include flint points and possible javelin heads as well as polished articles made of soft stone (slate and talc) and points made of bone.

## Giant Wolf

In 1995, a particularly interesting grave was discovered in the southern part of the Lokomotiv burial ground, on the right bank of the Angara River. The ground here contained burials from both of the above-mentioned Final Mesolithic periods.

BELOW Discovered at the Lokomotiv burial ground, this late Mesolithic joint-burial of a wolf and a man's skull is the first of its kind to come to light.

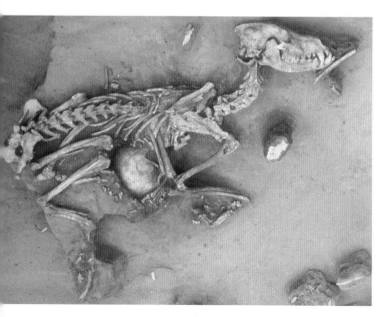

The burial of a tundra wolf (*Canis lupus albus*) was discovered in an oval grave pit with a north-south axis. The pit, filled with humus, was located approximately 3 feet (0.86 – 0.96 meters) below the present-day surface. The wolf was laid on its left side with its head raised in relation to its body. Between the wolf skeleton's elbow and knee joints, slightly below the chest, a man's skull was discovered, with the first and the second cervical vertebrae still connected to it. The skull was on an inclined plane with its base pointing downward. An oval spot painted in ocher was found to the west of the front paw bones. A young man's lower jaw, some finger bones, and several human vertebrae and ribs were also located on the same level as the wolf's bones, along the western side of the pit. The grave goods included two thin bone shanks, a fragment of a beaver canine, three bone needles and some stone blades and points. It turned out that the wolf had been buried in a preexisting grave. The lower jaw and the above-mentioned human bones are the remains of an Early Mesolithic burial, dating back to around 7,750 years ago. The grave goods may belong to that same burial. The wolf was buried 500 years later; the uncalibrated radiocarbon date from the wolf's bones is about 7,230 years ago.

The man's skull and the wolf may be interrelated and may have been buried at the same time. The color of the wolf's bones and of the skull is identical. The tundra wolf is one of the largest wolves. The body length of an average male is between 4 feet, 1 inch and 4 feet, 9 inches (1.26 and 1.45 meters), and its weight varies from 110 to 143 pounds (50 to 65 kilograms). The wolf was about nine years old, give or take a year, and it died in July or August. It remains a mystery how this cold-climate animal found itself thousands of miles away, in the south of the Baikal area of Siberia, given that this was the period of a climatic optimum (with high humidity and a warm climate). Its species is absent from the paleontological collections that have been put together in this territory. The wolf's teeth are slightly worn, which is not typical for its age, and may testify to the fact that the animal was domesticated.

Generally speaking, this ritual wolf burial is so far unique in archaeology. Many burials of dogs – but only of dogs – have been recorded from the Late Paleolithic period (in the north of Eastern Siberia), the Mesolithic (northern Europe) and the Neolithic (the Baikal area of Siberia).

## Early Neolithic Burials

No fewer than 222 human graves from the Early Mesolithic period, containing the remains of 304 individuals, have been found in the Baikal area of Siberia during the 120 years during which the graves have been

**ABOVE** An accumulation of grave goods from the burial ground. In addition to the various pieces of stone, bone and horn there are several lower jaws of small animals. They have been covered with a polished oval disk, and probably had either a military or cult purpose.

under study. The burial complexes are concentrated in the valleys of the Angara, Lena and Selenga rivers, as well as on the banks of Lake Baikal itself. On the basis of a large number of radiocarbon dates, their age has been pinpointed to between 6,000 and 7,000 years ago. The graves are dug into the earth, but in rare cases also feature stonework. The pits, dug into the base of a reddish-brown loamy soil, are between 1 foot and 8 feet, 2 inches (0.3 and 2.5 meters) deep. The bodies are mostly extended, supine and oriented to the northeast. All the burials are intensively painted with ocher. The graves contained individuals, couples and groups.

An approximately equal number of female and male burials, together with a small number of children's graves, were found in the graves for individuals and couples. The grave goods are usually richer in the male burials. In some of them, the objects, sometimes as many as 500, may have been packed in bags. The richest grave, in terms of the number of grave goods, was discovered in the Shamanka II burial ground on the southern bank of Lake Baikal. The woman's corpse, which had rotted considerably before it was buried (the skull was placed where the right shoulder should be, while the right shoulder bone was located in the pelvis area) was put into a pit, at the bottom of which various articles of bone, horn and stone – 292 articles in total – had been laid.

The burials of couples constitute a representative group of the Early Neolithic burial complexes found in the Baikal area. In the majority of cases, the burials were carried out simultaneously. Two ways of placing the bodies in graves can be seen: in the first, the bodies were laid in parallel, the heads pointing in the same direction; in the second, the bodies were laid in parallel, but the heads pointed in opposite directions. In most cases, the graves where the bodies point in the same direction contain two male bodies.

A grave opened at the Shamanka II burial ground contained two skeletons; one was in full anatomical order, while the other lacked its skull together with the first and second cervical vertebrae. A large accumulation of articles covered by an oval polished disk of serpentine

BELOW Early Neolithic joint burial of a man and woman. Both skulls were found to be wearing headgear. The man's was decorated with calcite, and the woman's has pendants of grooved wild boar's teeth.

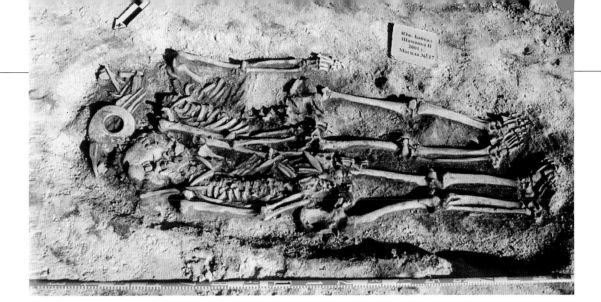

was located in the northeastern part of the grave pit, behind the first skeleton's skull. On the same skeleton's pelvis bones there was a bag with a variety of fishing implements. No grave goods were found next to the second skeleton.

Joint burials of men and women predominate in graves where the bodies are placed in opposite directions. The men are always mature or old, while the women are young. In the Lokomotiv burial ground, the male headgear was decorated with calcite rings, while the female headgear had pendants of grooved wild boar's teeth. No evidence of violent death has been found in the anthropological material.

Among the group burials known to us at present, no bodies were buried at different times. All the bodies were put into their respective graves simultaneously. With respect to the location of bodies, the graves can be divided into two groups: in the first, the bodies lie parallel in the horizontal plane, facing in the same or opposite directions; in the second, the bodies lie parallel in two or more tiers, pointing in the same or opposite directions. The graves of the former group mostly contain three bodies – those of a man, woman and child. The graves of the latter group contain both male and female bodies, most of them of a young age. Some evidence of violent death has been found in the graves belonging to the latter group.

The absence of skulls in approximately 12 percent of the skeletons is an intriguing peculiarity of these Final Mesolithic and Early Neolithic burial grounds. It is established that the heads were largely severed from the bodies between the second and third cervical vertebrae, but separate graves containing the skulls have not been found.

The people of this North Asian population were of Mongoloid craniological type. The average longevity in the Baikal area during this period varied from 30 to 32 years. Fish was predominant in the diet, as is proved by the grave goods (numerous fishhooks and ground slate fishing lures) as well as the results of chemical bone composition analyses.

ABOVE The Shamanka II burial ground, showing the joint burial of two men. Judging from the general context of the grave, one man was buried with head and the first and second cervical vertebrae missing.

BELOW The burial of a middle-aged woman at the Shamanka II burial ground, where the skull of the woman has been placed on the right side of her thorax.

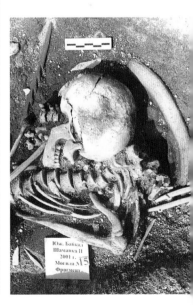

# A Woman from Roman London:
## In a Lead Coffin

## The Cemeteries of Roman London

During the redevelopment of the Spitalfields area of London in 1999, a Roman cemetery was discovered lying outside Bishopsgate, on the north side of the Roman city of Londinium. In the Roman period, it was usual to bury the dead outside settlements, and this cemetery flanked the road that became known as Ermine Street (which led to York and the north of England). Among the burials, which were dated from the third century AD onward, archaeologists uncovered a stone sarcophagus. When the lid was removed, it was found to contain an elaborately decorated lead coffin.

Such a burial was unusual for London; only two Roman lead coffins inside stone sarcophagi had been recovered from the city, both in the Victorian period. The practice of inhumation became more commonplace in Late Antiquity, especially with the rise of Christianity. Nearby, on the east side of Ermine Street, were the remains of other fragmentary stone sarcophagi, as well as traces of an elaborate wooden mausoleum that contained the remains of a child.

As the sarcophagus was removed from the surrounding clay, a number of grave offerings were found. These included glass containers, probably originally containing perfumed oil, as well as small containers made of jet, probably imported from near Whitby in Yorkshire, which perhaps originally contained cosmetics. One of the glass containers is of a shape rarely found in Britain, which hints that it may have been a special import. A further glass container (an unguentarium) was found placed between the lead coffin and the sarcophagus. Other items placed outside the sarcophagus include a jet hairpin as well as a jet ring, which may well have been used in the elaborate hairstyles of Roman women. Such offerings may indicate that the woman came from a pagan rather than a Christian family. The style of the objects pointed to a date between AD 350 and 375. This places the burial in the last phase of the province before it was officially abandoned by Rome in the early fifth century.

The sarcophagus containing the lead coffin was removed to the Museum of London so that it could be opened in sterile conditions. As the lid was removed, the team of

BELOW Conservators at the Museum of London cleaning the lid of the lead coffin before it was opened. The detail of the scallop shell pattern can be seen. The coffin is still in the stone sarcophagus.

conservators saw that the body of the individual had been preserved, in part due to a layer of damp silt. The coffin was cleared with extreme care, taking nearly a week.

The head of the woman had originally been placed on a pile of bay leaves. The body itself had probably been placed in the coffin in elaborate clothes, as traces of textiles were found lying underneath the body. Analysis showed that there had been a piece of silk decorated with gold thread, some 0.004 inch (0.1 millimeter) in diameter. Silk, derived from China, entered the Roman empire, probably in the form of thread, through the eastern provinces, most probably across the Syrian desert. An example of silk excavated at the frontier city of Dura-Europos has been shown to have been derived from northern China and probably dates to the Han period (206 BC–AD 220). Even earlier examples of silk are known from classical Greece. Fragments of gold-woven textiles, known from classical literary sources, are extremely rare finds in the archaeo-logical record. Other textile fragments from the Spitalfields burial were made of wool.

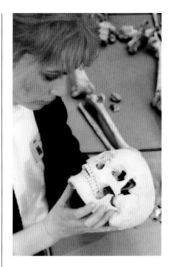

**ABOVE** The bones of the Spitalfields woman under examination.

## The Study of the Human Remains

The skeleton showed that the person buried inside the lead coffin was a young woman, probably in her early twenties. There were no apparent signs that she had ever given birth, so possibly she had been unmarried. Apart from the expensive clothes and the nature of the burial, two key details indicated that the woman had come from a relatively prosperous family. Firstly, at 5 feet, 4½ inches (164 centimeters) tall, she was above the average height for women in this period. Indeed, it seems that she was almost too tall for her coffin, suggesting that the coffin may not have been made specially for her. Second, a study of her teeth showed little sign of decay, and this may also indicate that she had access to a good diet. There was no indication from her bones about the cause of death; one possibility is that she succumbed to an infectious disease.

The analysis of the woman's teeth indicated something about her origins. In childhood, teeth absorb lead that is present, for example, in the drinking water. Once the enamel in the teeth has formed, the lead remains stable throughout life and in death. Lead has a number of isotopes, and each source of the metal has its own recognizable fingerprint. A study of the isotopes present showed a strong dissimilarity with the lead ores found in Britain, suggesting that the woman had spent her childhood elsewhere, possibly in the western Mediterranean in Spain, southern France or even Italy. It was possible to extract a sample of DNA from one of the woman's molars. The analysis showed a similarity with the profile expected from individuals living in the Basque region of Spain. As it would have been unusual for a single woman to have moved over such large distances, a likely explanation is that she moved to London with her family; perhaps her father was a wealthy merchant or one of the many administrators required to run the highly bureaucratized provinces.

ABOVE The bones of the woman lay on silt inside the coffin. Members of the museum staff had to wear protective clothing and masks to protect themselves from lead dust.

## The Lead Coffin

The lead coffin was decorated with scallop shells set inside a lozenge pattern. A parallel can be found in the lead coffin, placed inside an elaborately carved sarcophagus, that was found in the Minories, London, in 1853 (and which is now in the British Museum). A sample of the lead from the Spitalfields coffin underwent lead isotope analysis. The study indicated that the lead had been mined from the Mendips in Somerset. This source of lead had been exploited from the earliest days of the Roman invasion of Britain, and a mining settlement was established near the modern village of Charterhouse-sub-Mendip. Lead pigs (ingots) from those mines can be identified by the inscriptions they carried. More than 20 are known, mostly from the West Country and southern England. They have sometimes been found in pairs, suggesting that they were the load of a single pack animal. The latest dated lead pigs from the Mendips bear the name of the joint emperors Marcus Aurelius and Lucius Verus in the AD 160s; the honorific titles assigned to the emperors allow them to be dated in a narrow date range. However, the mines were probably exploited into the third and fourth centuries. The coffin was likely made in London.

The limestone – Barnack stone – used to form the sarcophagus had been imported from the East Midlands. Given that other burials in the Spitalfields cemetery had probably been placed in wooden coffins (traces of the nails have been found) or even just in shrouds, this choice of a stone that would have incurred expensive transport costs again points to the wealth of the woman's family.

## Reconstructing the Face

The excavation of the woman coincided with the making of a BBC television series, *Meet the Ancestors*, which explored human remains from newly excavated sites. As part of the coverage of the Bishopsgate excavation, the *Meet the Ancestors* team decided a reconstruction of the face would enhance the interpretation of the burial. The program makers approached a team from the Unit of Art in Medicine at Manchester University, which had been involved with the reconstruction of a number of faces, including that of the person buried in the royal tomb at Vergina in northern Greece.

The Spitalfields woman was one of a number of attempted facial reconstructions from Roman Britain. Among the first, in 1991, was of a man, aged between 45 and 50, buried in a chalk-packed lead coffin, decorated like the Spitalfields coffin with scallop shells, in St. Stephen's Cemetery along Watling Street at Verulamium (St. Albans) and probably dating to the third century AD. A second commission was of a middle-aged woman, probably dating to AD 330–350, buried in a wooden coffin at Camulodunum (Colchester).

The same BBC team had also been involved with the reconstruction of a Roman man, aged around 45, buried, along with the bones of a woman, in a stone sarcophagus at Mangotsfield outside Bristol. In this case, a lead isotope study had shown similarity with the Mendips field, which indicated that the individuals had grown up in the locality. A further body was recovered from a lead coffin, originally placed inside a wooden outer coffin, at Venta (Winchester). The burial could be dated because the dead individual had been given a coin minted in AD 316 or 317 (in the reign of the emperor Constantine) to pay to enter the underworld. A study of the bones identified the individual as a male. Like the Spitalfields woman, the Winchester man's teeth showed little sign of decay, again indicating the likelihood of a relatively high social status. This is one of several bodies buried in a lead coffin from the same city.

The detail of the Spitalfields woman's reconstruction was informed by fourth-century representations of women, in particular from portrait sculpture. This was particularly important for the way to represent the hair, which in the reconstruction was pulled to the back of the head. The reconstruction now forms part of the display at the Museum of London.

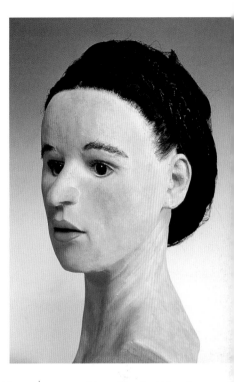

**ABOVE** A team from the University of Manchester were able to reconstruct the woman's face. The hairstyle is based on representations known from contemporary sculpture.

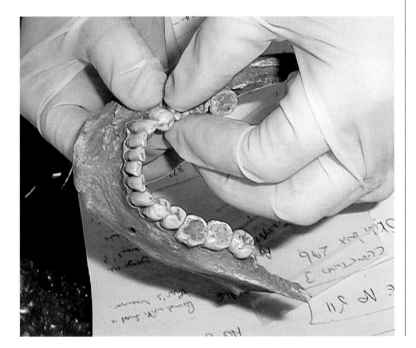

**LEFT** The teeth of the woman were in good condition suggesting that she had a good diet. Analysis of the molar teeth showed that she had come from Spain.

# The Romito Dwarf

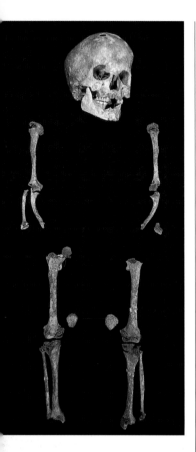

ABOVE The skeleton of the Romito adolescent, the earliest known example of a dwarf in the world.

BELOW A general view of the Riparo del Romito, located at the base of the rocky hill in the foreground.

The Riparo del Romito, a large rock shelter in the province of Cosenza, close to the town of Papasidero (northern Calabria, Italy), was excavated in the 1960s by the late Paolo Graziosi, the great Italian rock art specialist. The front, outer part of the site, under the rock overhang, contained occupation material of the late Upper Paleolithic, starting at about 18,750 years ago. It also features a remarkable rock on which three aurochs (wild cattle) had been deeply engraved – one large bull, nearly 4 feet (1.2 meters) long, and two smaller ones beneath.

During Graziosi's excavations, six paleolithic skeletons were unearthed in four burials, as were isolated human bones. The most remarkable grave was one that contained two skeletons, known as Romito 1 and 2, unearthed in 1963. They are of a middle-aged female and an adolescent male who was a dwarf – the earliest known example of a dwarf anywhere in the world.

The two skeletons had been buried in a shallow, oval pit. Although it is not totally clear whether they were interred at the same time, as they were found one above the other, a simultaneous burial does seem likely; the female (Romito 1) seemed to be clasping the dwarf (Romito 2) with her arm, while the back of his neck rested on her cheek.

Unfortunately, the exact disposition of the remains will never be known, as no drawings or photographs were ever published. The archaeological level in which this grave occurred has been dated to 11,150 years ago. Two large fragments of horn were found with the bodies, presumably as grave goods, as similar objects occur in other Italian graves of this period.

It was surprising, to say the least, for a dwarf to be encountered in such remote times, as the oldest specimen known previously was from predynastic Egypt, some 5,000 years later, and the condition is rare even in large populations. In general, dwarfism results from abnormalities in cartilage growth that lead to a greatly reduced size of the long bones. Romito 2 displays a dwarfism syndrome known as *acromesomelic dysplasia* in which there is an extreme reduction of the radius and ulna and a shortening of the hands and feet.

The specimen, now housed in the Museo Archeologico Nazionale in Reggio Calabria, comprises a virtually intact skull, together with the long bones. The absence of the pelvis makes sexing difficult, particularly in an abnormal individual like this, but it seems to be a

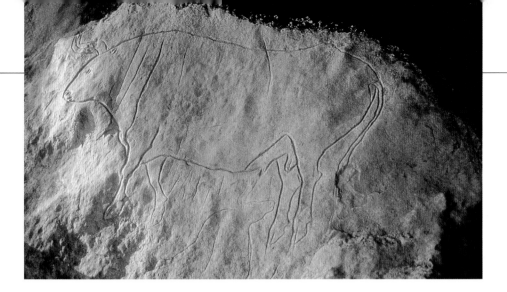

male on the basis of several features of robustness in muscle attachments. Aging is also difficult, but various aspects of bone growth and tooth development (the wisdom teeth are not yet fully erupted) point to an individual between 14 and 19 years old, most likely around 17. There is no trace of trauma or violence.

The skull has pronounced frontal and parietal bossing, including most of the forehead, which is very high and extremely flat across the front. The cranium as a whole has a domed, pentagonal shape when seen from above, and it displays an infantile facial architecture that is small compared to normal individuals from the late Upper Paleolithic period, though tooth size is within the normal range.

Each of the long bones is remarkable for its short length, especially the radius and ulna, which are the most bowed and deformed of all. Overall, the limbs are about half normal size. The dwarf, although not yet fully an adult, would probably not have grown any further, as all the epiphyses of its long bones were fused. Its height has been estimated at between 39 and 51 inches, or 3 feet, 3 inches and 4 feet, 3 inches (100 and 130 centimeters), probably about 50 inches (127 centimeters).

Romito 1, on the other hand, was a small but normally proportioned individual, most probably a female and somewhat older at death than the dwarf. The skeleton is less complete than that of Romito 2, but this individual was probably about 57 inches, or 4 feet, 9 inches (144 centimeters) tall, one of the smallest and most gracile people ever recovered from the Upper Paleolithic of Europe. Her tooth wear and degree of bone fusion suggest that she was at least in her mid-thirties, more likely between 40 and 50 – in other words, at least 20 years older than the dwarf.

It is therefore possible that this double grave contained a mother and her dwarf son. Despite the difficulty of establishing a familial relationship (in the absence, so far, of DNA tests), the two skeletons do display some minor anatomical features in their skulls and teeth that support some kind of link. The age difference makes it unlikely that they were husband and wife, though this cannot be ruled out entirely, particularly as both were of short stature.

**ABOVE** The great aurochs figure, 4 feet (1.2 meters) long, engraved on a rock in front of the Riparo del Romito; a second, smaller aurochs figure can be seen beneath its belly, its head touching the back leg.

**BELOW** The dwarf's ulna (elbow/forearm bone) compared to that of a person of normal growth. His inability to fully extend his elbows must have impeded his full participation in the hunt.

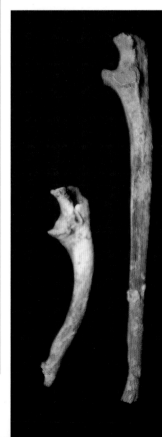

Be that as it may, the really important point to emerge from the Romito grave is that Upper Paleolithic society displayed care and affection for people with handicaps. Although we have other examples showing care for normal people made infirm by trauma and/or age (such as an elderly Neanderthal at Shanidar, Iraq), Romito 2 is an individual whose disease was genetic in origin and had a radical effect on his appearance throughout his life. In all likelihood, thanks to low population density, his social group had no previous experience of such a rare condition – an infant of abnormal growth who could not walk properly – although this syndrome does suggest some degree of inbreeding.

One can speculate that his condition must have been an impediment to survival, as although he probably had normal intelligence and was free of serious medical problems, he could presumably not have contributed greatly to the group's economic life. Dwarfs usually have a waddling gait and, although capable of rapid movement, their legs and back tire after walking short distances. In view of the rugged terrain in Calabria and the probability that Upper Paleolithic groups in this region migrated seasonally, life cannot have been easy for this youth. His limited ability to extend his elbows to more than 130 degrees was a further impediment to his effective participation in hunting activities. Dwarfs of this kind are usually susceptible to back problems and joint pain.

In short, Romito 2 had formidable physical handicaps for the life of a nomadic hunter-gatherer of the Ice Age, and the fact that despite them he survived to late adolescence must mean that his social group protected and supported him. Indeed, it is likely that his burial in this special place, in front of some rock art (as in the case of the Cap Blanc Lady, p. 108), testifies to his important social status.

# Anne Mowbray and the
## Skeletons in the Tower

When the bones of two children were found during restoration work at the Tower of London in 1674, it was speculated that they could be the sons of Edward IV, who had disappeared from the Tower shortly after their father's death in 1483. This possibility would be strengthened if their respective ages fitted those of the princes – fortunately, enough of the jaws and teeth survived for an age assessment to be made – and also if evidence of kinship with known family members were found; the remains of two of their relatives, Anne Mowbray and Mary of Burgundy (Marie de Bourgogne), have been described.

## Dental Age

Teeth develop lengthwise from crown to root, growing at a regular rate until they are ready to erupt into the mouth. Eruption times vary among individuals, but the stage of development of any tooth is closely related to age. Each tooth's stage of development, whether it is loose or seen in a radiograph, is therefore evaluated and dental age inferred. Sometimes the developmental stage of front teeth lags behind that of back teeth, or deciduous teeth lag behind permanent teeth. A well-established score system gives a value to each stage of development. The value for each tooth is added to yield a maturity score that relates to the age of the child.

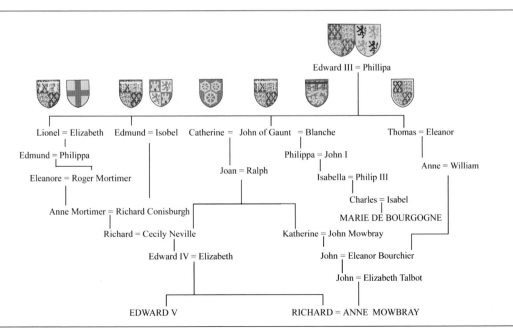

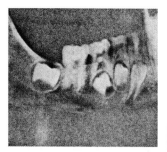

**ABOVE** Radiograph of Anne Mowbray's jaw, showing developing teeth.

Does the method work for medieval children? We could test it on Anne Mowbray, who died aged 8 years, 11 months in 1481. Her teeth were studied in 1965 by Martin Rushton, who published radiographs of her jaws from which we can assess her teeth. Both deciduous and permanent teeth were present. Only the permanent teeth were analyzed. From the radiographs, each of the permanent teeth can be given a score. The resulting maturity score equates to an age between 7 years, 9 months and 9 years, 3 months. This neatly brackets Anne's known age at death.

## Dental Age of the Skeletons in the Tower

Although only three teeth survive in the lower jaw of one of the Tower skeletons and none at all in the other, we can still assess the developmental stage of the teeth from the shape of the bony sockets. For the skeleton with surviving teeth, a maturity score can be calculated that gives an age between 8.5 and 10.75 years, the average of which is 9.5 years old. The bony sockets of the other skeleton indicate a maturity score of at least 12.7 years, which is more than two and a half years older than the first child. Edward IV's older son was born on November 2, 1470, and his brother Richard on August 17, 1473 – 2 years, 9 months later. Can this be mere coincidence?

**BELOW** Front view of the skull of Anne Mowbray, 1472-1481.

## Relationship

Alan Brook, who studies the prevalence of congenitally missing teeth (known as *hypodontia* or *agenesis*), records a prevalence of only 1 percent missing teeth, other than third molars, in the population as a whole; but the incidence increases to 30 percent in close relatives. Congenital absence of teeth runs in families, although not always the same teeth are missing.

When the jaws of the two skeletons were first examined, the X-rays showed that the upper premolars of the older were not present and had never developed, nor had the lower third molars. Looking at the dental chart for Anne Mowbray, Rushton noted a rare form of hypodontia. She was missing her second molars on one side of both the upper and lower jaws. Anne, when only six years old, was married to Richard, the younger of the princes. A special dispensation was needed for this marriage because the two were closely related cousins. The tomb effigy of Mary of Burgundy (1457–1482) in Bruges, Belgium, shows that she had an unusual upper lip. When French tooth specialist Pierre-

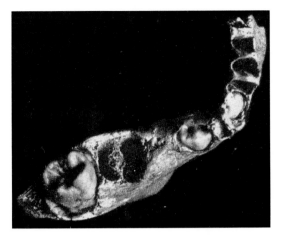

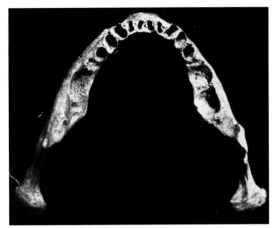

François Puech matched her skull with the head on her tomb, he realized that her unusual lip form occurred because she was missing her upper lateral incisors. She was descended from Edward III through her great-grandmother, Philippa, a daughter of John of Gaunt. Her parents were first cousins. The chance that other descendants of Edward III, including the children of Edward IV, would have dental hypodontia is much greater than 1 percent.

There is another hint of a relationship between Anne Mowbray and the skeletons from the Tower. Small bones between the main bones of her skull are similar to extrasutural bones on the skulls of the two skeletons.

**ABOVE LEFT** Lower jaw of the younger of the two skeletons found in the Tower of London.

**ABOVE RIGHT** Lower jaw of the older child in the Tower.

## Consanguinity among the Descendants of Edward III

The sons of Edward IV (Edward V and Richard, the princes who disappeared in the Tower), Anne Mowbray and Mary of Burgundy were all descended from the sons of Edward III: Lionel, John of Gaunt, Edmund and Thomas. Edward III was Anne's great x 4 grandfather

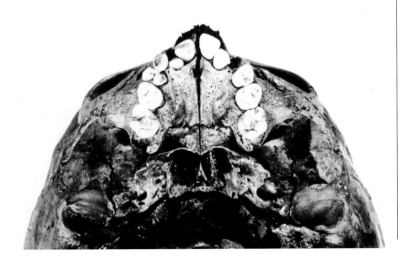

**LEFT** Anne Mowbray's teeth were studied in detail by Martin Rushton in 1965.

RIGHT Side view of the skull of Anne Mowbray. There are extra small bones between the main bones of the skull.

ABOVE Skull of Mary of Burgundy showing the space for the undeveloped incisors.

through her great-grandmother, Joan, a daughter of John of Gaunt; he was also her great x 3 grandfather through her great-grandmother, Anne, daughter of Thomas.

Similarly, Edward III was Mary's great x 3 grandfather through her great-grandmother, Philippa, daughter of John of Gaunt. Edward III was also the princes' great x 3 grandfather through their great-grandmother, Joan, a daughter of John of Gaunt, and also through their great-grandfather, Richard, son of Edmund. He was also their great x 5 grandfather through their great x 3 grandmother, Philippa, daughter of Lionel. John of Gaunt was ancestor to all four. Thomas Neville was Anne's maternal great-grandfather; his brother Ralph was the princes' great-grandfather.

Anne was quite small for her age, perhaps only 4 feet, 4 inches (133 centimeters) tall. The two skeletons were also small, especially the older of the two. But this is a general finding for children from medieval times.

## When Did the Children Die?

If the evidence for consanguinity can be accepted and the skeletons found in the Tower in 1674 are indeed those of Edward V and his brother Richard, Duke of York, younger by 2 years, 9 months, and one wishes to propose that they died at the same time, the most likely period that is compatible with the dental age for both skeletons would be some time in the year 1484.

# The Mysterious Burials of the
## Okunev Culture

A hundred years ago in the steppes of South Siberia, one could still see endless stone monuments: stelae and statues, short and tall – up to 13 feet (4 meters) – made of sandstone and granite, standing on burial mounds or separately. At the hours of sunrise and sunset, with oblique lighting, engraved images became visible on many of them: the features of anthropomorphous faces, decorated with lines and curves, with three eyes and a big mouth in relief, often with animal horns and ears, and with a complicated headdress. Some figures with a big stomach and breasts probably represented pregnant women; some faces were surrounded by solar rays; a few looked like realistic human faces. The design of the statues often combined relief carving with engraving on the flat surfaces, and sometimes traces of red ocher are preserved in the lines. One can imagine what a strong impression these fantastic decorated face-masks must have made on their contemporaries, since in the eighteenth and nineteenth centuries those statues, already worn by weather and time, still provoked fear and worship among the local Siberian peoples.

ABOVE Many of the stone stelae were decorated with engraved images that only became visible at certain times of the day, typically sunrise or sunset, because of the way the natural daylight fell on the rock's surface.

The unusual look of these steppe idols and the legends surrounding them attracted the attention of the explorers of the Yenisei steppes. They were first described in the works of Daniil-Gotlieb Messer-schmidt, who was sent in 1719 by Peter the Great to study the nature and population of Siberia. Since then they have always been, and still are, the subject of unremitting scholarly interest. In the late nineteenth and twentieth centuries, most of them were taken to museums. Yet, despite the long history of these monuments, their cultural attribution was only established quite recently, when in the 1960s an absolutely new culture of the Early Bronze Age was identified by the Russian archaeologist Gleb Maksimenkov from the results of excavations in several cemeteries. The culture was named Okunev after the village of that name in Khakassia, where the first graves of this type were excavated. In the cemeteries, the excavators found stone slabs decorated with images of the same face-masks that were known on the stone statues. Moreover, some features of the funeral rites also made it possible to connect the famous monuments with the burials of the Okunev culture.

This culture, one of the most dazzling and mysterious of the Bronze

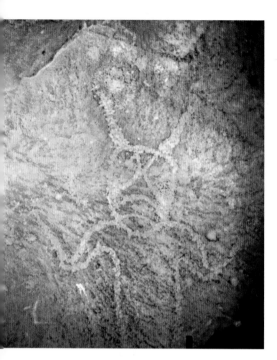

Age, flourished in the first half of the second millennium BC in the Minusinsk hollow of South Siberia, an area of forest-steppe in the basin of the Yenisei River. It emerged from the symbiosis of the cultural traditions of the local Neolithic population and those of alien groups of cattle breeders. The combination of these heterogeneous elements produced the peculiarity of the Okunev culture.

The culture is incredibly rich in art, and the versatility of imagination expressed by its artists – and their artistic talents as well – is amazing. They left stone statues, decorated slabs, engravings and paintings on rocks, miniature sculptures, carvings on bone and a multitude of decorated objects. One can only guess how much art executed in organic materials has not survived. However, despite the diversity of the Okunev art imagery, and perhaps because of it, much about these people remains obscure. Their funerary rituals are even more mysterious.

## Burial Variations

**ABOVE** Ocher played a significant role in the funerary rituals of Okunev society, and traces can be found in paintings and engravings on the stone burial slabs.

**BELOW** The stone statues often combined relief carving with engraved images and betrayed a culture rich in art.

On the modern ground surface, the Okunev burial mounds (kurgans) are barely visible, and nearly all were found accidentally when their perimeter slabs became visible. The cemeteries consist of 2 to 14 kurgans, their shape being a regular square with sides from 16½ to 33 feet (5 to 10 meters); the biggest has sides of 131 feet (40 meters). The square is defined by sandstone slabs vertically dug into the earth around the perimeter. Inside the square are several graves, constructed in a box shape made of stone slabs (often decorated with pecked, engraved or painted images). The men's boxes range in length from 3 feet, 7 inches to 5 feet, 7 inches (110 to 170 centimeters) and in width from 1 foot, 4 inches to nearly 3 feet (40 to 90 centimeters). Women's boxes were smaller, while children's boxes were less than 3 feet, 3 inches (1 meter) in length. Graves with several bodies did not differ in size. The boxes are 1 foot, 4 inches to 2 feet (40 to 60 centimeters) deep.

Such small tombs were required by the method of burial. The bodies were always placed in the grave on their back with the knees lifted high and acutely bent. The graves had a floor paved with small stone slabs, sometimes slightly raised at the head end, and a ceiling made of one massive stone slab or several smaller ones. Some boxes contain so-called pillows – a rock support for the head and shoulders, or a vertical slab with a groove for the head. Several burials were found in pit graves and in niches in the side wall of a shallow shaft.

Individual burials of a man or a woman are prevalent, but burials of a couple, a woman with a child or children, and a couple with a child or children have also been found. It has been ascertained that the burials of pairs were carried out at the same time, with the bodies lying close to each other in a small box; if there was a child, it was placed on the female skeleton. It is not quite clear who these men, women and children buried together at the same time were. Sometimes they cannot be actual parents with their children because of their ages. They may just be adults and children who belonged to a group and who died of the same disease.

There are also graves with burials that clearly took place at different times; the initial skeleton was taken out, a new body placed inside, and the old bones carefully replaced. Perhaps this happened in the winter, when it was impossible to dig a new grave. Individual child burials never occur within the stone perimeters; children were buried either with adults or in old, nonfunctioning cemeteries. The pediatric death rate in Okunev society was quite high; children make up half of all the deceased.

One of the strangest features of the Okunev funerary ritual is the burial of some bodies without a head. The heads were probably kept somewhere until the death of their keeper. There are several graves of such keepers where, in addition to a complete skeleton, between three and eight human skulls without jaw bones were found. They are definitely not the skulls of enemies, as the beheaded bodies are buried nearby. There are also rare but intriguing cases where disarticulated human bodies (with the parts of the skeleton found in anatomical order, suggesting that partial bodies had been buried), or just bones were found. One grave contained the bones of an elderly man not in anatomical order and gnawed by a dog or wolf.

The burials were most likely carried out in the warm season, so the living somehow had to try to preserve the bodies of the deceased through the winter until they could be buried. One of their methods could be the

ABOVE Grave goods in male burials were rare and are thought to be associated with men of considerable importance – perhaps a shaman, a chief or a priest. Often, the items found would be quite unusual, such as this decorated egg.

LEFT Burial practices of the Okunev people have mystified archaeologists, with a number of unexplained group burials. Here, one skeleton has been buried with as many as six skulls – all missing their jaw bones.

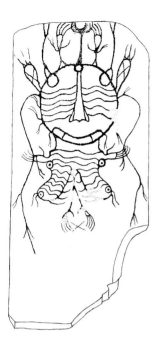

post-mortem trepanation of the skulls for removal of the brain. Trepanated skulls are found in only one of the Okunev cemeteries, which dates to the later stage of the culture.

Most of the Okunev skulls are characterized by some slanting and flatness in the occipital-sincipital area. Anthropologists consider this an artificial deformation of the skull, deliberately produced by keeping a baby in a hard, short cradle for a long time and with something like a bag of sand instead of a pillow (such cases are known from ethnography). This feature is even reflected in female steatite figurines, which shows how much attention the Okunevians paid to the shape of their heads.

## Grave Goods

The deceased were buried in clothes decorated with animal canine teeth and lots of beads. Women also had decorations on their hair, ears and neck. Their shoes were embroidered with sable teeth (from 70 to 160 teeth on each shoe!). The indispensable item for women in this culture was a needle case of bone or bronze. Women and children also received ceramic vessels containing food and drinks. The men's burials had no grave goods at all, with the exception of several cases that have been named the "unusual Okunev burials." In these few cases, men were buried with a great variety of unusual objects such as marble balls with a drilled hole; bronze and stone axes without any traces of use; bone knives; the skull and beak of a crane, the penis bone of a bear, metatarsals of wolves and roe deer and other strange bones; a horn staff with the carved head of a predator at the end, with gaping jaws and long canines; a horn rhyton decorated with carved animal ears and horns and covered with ocher inside; and many other objects that are hard to describe and whose purpose is absolutely incomprehensible. These men sometimes also had unusual ceramic vessels in the shape of a low bowl on a ringed foot or on four feet. These were probably incense cups, as their insides have traces of fire and a compartment, perhaps for incense or aromatic seeds. These extraordinary assemblages have some scholars proposing that these are the graves of a special category of people. What role did they play in Okunev society? Were they shamans, chiefs or

RIGHT This skull shows clearly how ocher painted on the skin of the deceased has bled through to the bone beneath.

priests? It is clear only that they were somehow related to cult and ritual activity.

Another special group of finds comprises female images. They are either miniature sculptured heads, 0.6 to 2 inches (1.5 to 5 centimeters) long and carved in steatite, or engraved faces with long hair on flat bone plates, 1½ to 3½ inches (4 to 9 centimeters) long. The images occur only on the upper part of these objects, while their lower part was probably inserted into something like doll's clothes. They have always been found in groups in the graves of female teenagers and are considered to be doll-idols, receptacles for the souls of deceased female ancestors. These dolls were doubtless inherited through the female line and after the death of the last heiress were buried with her.

Finally, one should note the special importance of ocher in the funerary rituals and beliefs of the Okunevians. There are ocher paintings, and engravings painted with ocher, on the decorated stone slabs used for the burial constructions. Ocher is also found in carved lines on female figurines and bone plates. The lines of the depictions on the stone monuments were also once painted with ocher. In the unusual burials, all the ritual objects and vessels were densely covered with ocher inside, and a vessel containing pieces of ocher, a stone slab, and a little pestle for grinding paint was found. But the main peculiarity of the funerary ritual was that the bodies and faces were probably painted with ocher in the same way as the fantastic creatures and face-masks on the stone statues and slabs. Traces of ocher were found on some bones, and one can clearly see the characteristic ocher lines on some skulls , exactly like the ones on the face-masks, that transferred to the skull bones from decayed skin. On the rocks are painted images thought to be schemas or patterns for facial decoration. Probably, special individuals in Okunev society had the job of decorating bodies for funerals and other rituals in accordance with canons and the highly developed Okunev mythology, which are also imprinted in the beautiful monuments of Okunev art.

**BELOW** It was not unusual for the stelae to be as tall as 13 feet (4 meters), and one can easily imagine the impression they must have made on the Siberian land-scape, their engravings inspiring awe and wonder.

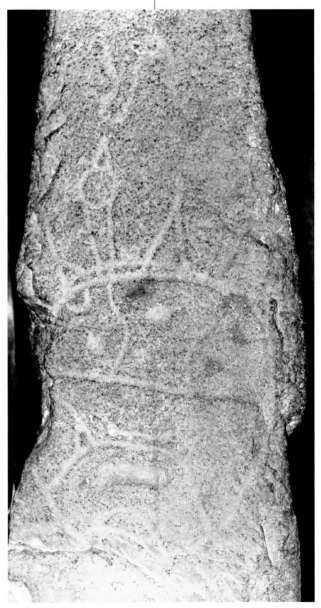

# Positioned for Political Influence

RIGHT The Maya ceremonial center of Chichen Itza in central Yucatan, Mexico, showing El Castillo, a square-based stepped pyramid, c. 82 feet (25 meters) high. In the foreground is a "chacmool", a reclining human figure used as an altar for offerings.

L ife in the fifteenth century was full of surprises for European explorers. But when Christopher Columbus, on his fourth voyage to the Americas, spotted and boarded a large ocean-going cargo canoe, he marveled at the inventory of trade items, which included sacks of chocolate beans, cotton cloth and copper bells. Columbus encountered the canoe in the Bay of Honduras off the Caribbean coast of Central America; the paddlers spoke a language that was completely unknown to Europeans: Mayan. These fifteenth-century Maya merchants had inherited the legacy of one of the great civilizations of the Americas.

Classic Maya society (AD 250–900) once flourished in the jungles of southern Mexico and northern Central America, where a productive system of tropical farming supported a population of several million people who were governed by divine rulers. As is the case with all ancient civilizations, their monumental constructions – towering pyramids and palatial residences – now stand in ruins. But contrary to popular understanding, the Maya have not vanished. Over five million people living in southern Mexico and the northern part of Central America still speak one of the many Mayan languages. Impoverished and politically disenfranchised by 500 years of subjugation, contemporary Maya people struggle to regain both a political voice and the lands of their ancestors.

Maya roots penetrate deeply into the past in this part of the world; recognizable Maya culture can be identified by archaeologists as early as 1200 BC. During the ensuing Late Preclassic and Classic periods (400 BC–AD 900), emphasis on the construction of monumental architecture and realistic royal portraiture resulted in some of the most impressive and aesthetically appealing archaeological sites in the world. For this reason and many others, Maya archaeology holds great fascination for scholars and laypeople alike.

During the Classic period, a strong literary tradition developed among royal Maya court societies. Ongoing decipherment of hieroglyphic texts, both painted and carved, provides us with the names of rulers and details of their royal activities, further personalizing a connection with the past. Finally, the Maya belief system included ancestor veneration, and continued importance was accorded to powerful, influential individuals, even after their death. This belief was expressed in elaborate mortuary rituals among commoners as well as royalty. The material remains of these rituals provide archaeologists with a wealth of information regarding Maya death ways.

## Interpreting Burial Positions

Pre-Columbian Maya funerary customs belie the old adage, "you can't take it with you." Divine rulers were often buried with necklaces, earspools and death masks crafted from precious jade stone as well as elaborately painted chocolate-drinking vessels that frequently bore a hieroglyphic "monogram." During certain periods, the anatomical position in which deceased Maya individuals were interred conveyed a potent message of authority and status. This seated position is evocative of the posture in which rulers were portrayed – usually within a throne room – on Classic Maya painted pottery. A scene from the Madrid Codex (one of four extant fan-fold, hieroglyphic books dating to the Postclassic period, AD 900–1500) shows the manner in which seated burials were tightly wrapped in strips of cotton cloth and provided with ritual offerings of food.

**BELOW** Maya ruins of monumental constructions date back to at least 400 BC and are epitomized by towering pyramids and palaces.

Despite the rapidity with which a corpse deteriorates in a tropical environment such as the Maya Lowlands, this image, painted by an unknown Maya scribe, suggests the possibility that wrapped and seated corpses may have been displayed above ground for a time before receiving a subterranean burial. In fact, many seated burials have been excavated archaeologically from sites dating to the Postclassic period.

Ironically, deceased members of Maya society were rarely interred in a seated position during the Classic period; instead, rulers and commoners alike were more likely to be buried in an extended supine position. Regardless of their "relaxed" position, powerful yet deceased members of Classic Maya society appear to have continued to exert considerable political influence from the afterlife. Both archaeological and hieroglyphic sources indicate that royal tombs, in particular, were repeatedly reopened so that the living could consult with ancestral figures of authority. At Copán, the royal tomb of a female buried within a structure referred to as "Margarita", reveals evidence of a later visit, at which time a red mercury-based powder (cinnabar) was sprinkled over her decomposing bones. In all likelihood, the purpose of such visits was linked to reaffirmation of the legitimacy of the current ruling dynasty.

The mortuary practice of preparing a corpse for burial in a seated position is rooted in the earlier Preclassic or Formative period (1000 BC–AD 250), particularly the latest part of the Preclassic period (400 BC–AD 250), when deceased males, in particular, were interred in a cross-legged position. Without a doubt, this time was one of pronounced political centralization as seats of power, many of which endured through the Classic period, emerged throughout the Maya Lowlands. On the eastern side of the Lowlands, in the northern portion of a small Caribbean nation known today as Belize, Maya settlements proliferated during the Late Preclassic period. The combination of rich soils and the diversified protein resources of a riverine-wetland environment provided a stable subsistence base for maize-based farmers. Archaeological sites such as Lamanai, Colha, Cuello, Cerros and K'axob supported growing populations, and the demarcation of status and power, both in life and death, was accentuated increasingly over time.

Nowhere is this pattern more clearly seen than at the site of K'axob, where over 100 Preclassic burials were encountered and excavated during a long-term project directed by the author and focused on gaining an understanding of the genesis of ancestor veneration. K'axob is situated between a large perennial wetland and the New River; the latter flows leisurely to the north and discharges into the Caribbean Sea. Living in an environment rich in aquatic resources such as fish and turtle, K'axob

**ABOVE** Among the many faces carved in stone (the same stone used to build the monumental structures) are images of living people as well as representations of characters from the underworld.

farmers were well fed, and their skeletal remains suggest that they rarely suffered from frailties linked to the protein shortages common elsewhere in the Lowlands.

As was customary in Pre-Columbian Maya society, only selected deceased members of a family were accorded an honored place of burial. Excluding royal burials of the Classic period, the place of interment was likely to be a prepared pit that had been dug through the floor of a domicile or into a plaza surface that linked several residences. Due to the close association of residence with ancestral interment, household archaeology in the Maya lowlands is inevitably the archaeology of mortuary ritual.

When the village of K'axob was founded during the middle part of the Preclassic period, only extended burials were interred within the early domiciles. Somewhere around 200 BC, mortuary practices changed, and certain individuals were prepared for the afterlife by being wrapped in a cross-legged or tightly flexed position. The trend is quite pronounced, and there is little doubt that most corpses interred in this manner once held positions of authority within the community. By way of clarification, seated burials frequently contained distinctive mortuary accoutrements. Most of the red-slipped spouted pottery vessels, linked with Preclassic chocolate-drinking, were found in the burial pits of seated individuals. Many of the containers are signature, one-of-a-kind fabrications featuring elaborate modeled or incised decorations.

Large, flared-rim pottery bowls were also found with seated burials. In palace scenes painted on Classic Maya pottery, such bowls are filled with tamales, a food that traditionally formed an important component of Maya feasting cuisine. At K'axob, the bases of these bowls were often painted with a large cross, an indicator of centrality in Maya cosmology. Burial 1-1 provides an example of a seated individual buried at K'axob with two large bowls, both of which were painted with a cross motif.

**ABOVE** Pyramidal Temple 21 of Tikal, Guatamala, was built to honor the memory of a great Late Classic ruler, Jasaw Chan K'awil whose remains were uncovered at the base of the pyramid.

Although the body of the central seated person in Burial 1-1 was neither sprinkled with hematite (a red iron-based powder) nor adorned with jade, other seated burials of K'axob did receive such treatment. Initially, seated burials occurred singly, but increasingly, as time went on, seated individuals were accompanied by the secondary remains of others, as can be seen in Burial 1-1. In some instances, archaeological stratigraphy provided clear evidence that the burial pit had been reopened, either to conduct a commemorative ritual or to deposit the remains of additional, possibly related, individuals, or for both purposes.

In the case of Burial 1-1, the partial remains of six others – both male and female, with age estimates from young to old adult – were placed beneath, beside and above the primary seated interment. Such partial remains are called secondary burials because they generally include the remains of an individual whose body was either initially interred or defleshed elsewhere. Frequently, only the skull and long bones are present in a secondary burial, and often they are positioned as if originally deposited as a bundle. Close osteological inspection also tends to reveal that the bundled bones are more highly weathered than those of the primary, articulated skeleton, suggesting the possibility that the bundled remains were ancestral to the primary interment. At K'axob, the overall poor state of bone preservation (due to the wet tropical environment) combined with the sheer paucity of anatomical elements present in a bundle burial precludes determination of the cause of death – that is, whether death was inflicted, as in human sacrifice, or resulted from natural causes.

Although persons buried in a seated position could be either male or female, more males were buried in this fashion. Surprisingly, several young males of warrior age were accorded a seated burial, including the central figure in Burial 1-1. These young men could not yet have distinguished themselves in political leadership, but they may have proven themselves on the battlefield. The preferential manner of their burial hints at the increasing importance of military combat within Maya society. In short, the anatomical position of human skeletons, as encountered and recorded by archaeologists, can be "read" for information relevant to the status and political power of the deceased – who, in some cases, were positioned in death based on the political influence they exerted during their life. Given the significance of ancestors within traditional Maya society, it is likely that many of those who held positions of authority during their lives continued to flex political muscle long after their death.

**BELOW** This was a culture rich in literary and artistic tradition, and has left many realistic carved and painted royal portraits.

CHAPTER FIVE

# mummies and mummification

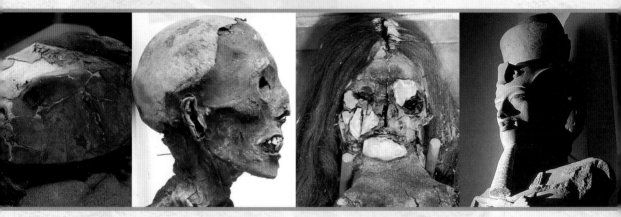

The careful preservation of human bodies shows that various cultures clearly believed in an afterlife. Most of the best-preserved bodies that have been discovered are the mummies of Egypt and South America, but there are also examples from other peoples of the past.

# Chinchorro Mummies

W hen most people hear the word *mummy* they think of King Tutankhamen and the other elaborately prepared Egyptian mummies that have toured the world in recent years. Such mummies were made by the highly trained mortuary specialists of ancient Egypt, one of the most complex and influential civilizations of the ancient world. The earliest known artificially prepared mummies, however, were actually prepared by members of small-scale hunting-gathering-fishing groups that lived half a world away and thousands of years earlier – the Chinchorro people of coastal northern Chile. These remarkable mummies, some as much as 7,000 years old, are still being discovered in and around the town of Arica, Chile.

Chinchorro mummies were reported in the Arica area during the last quarter of the nineteenth century, beginning just after the War of the Pacific. The first accounts were mainly in newspapers, and the mummies were discussed at an International Congress of Americanists at the beginning of the twentieth century. Max Uhle, one of the pioneers of Andean archaeology, described the numerous mummies excavated from the base of the hill of El Morro in Arica and from the Chinchorro site on the beach. Uhle estimated that the mummies were less than 2,000 years old. Only in the late twentieth century, after the development of

BELOW The clay masks found on many Chinchorro mummies were often painted black or red.

radiocarbon dating, did archaeologists discover that the Chinchorro mummies were perhaps 7,000 years old.

The Chinchorro tradition of mortuary treatment first appeared thousands of years ago on the arid coast of what is now northern Chile and far southern Peru. The most elaborate Chinchorro mortuary treatment dates to between roughly 4,000 and 7,000 years ago (5000 to 2000 BC). The mummies were made in a variety of ways. Some bodies were simply wrapped tightly to form stiff bundles that mummified naturally in the dry climate of the Atacama desert. Others were covered with local clay and with black or red pigment. Some had specially modeled clay masks and clay models of genitalia (in the case of males) or breasts (in the case of females). Still others were completely artificially constructed. In these cases, the skin, soft tissues and organs were removed, bones were reinforced by being wrapped and tied to sticks and the shape of the body filled in with vegetation. Then the skin was replaced and the whole mummy covered with clay. Each mummy is unique, but from the outside, they all have a recognizably human form.

Chinchorro mummies are most often found in groups, sometimes in cemeteries up to 66 to 98 feet (20 to 30 meters) across, sometimes in smaller groups at the settlements where people lived. In many cases, the groups of mummies include both adults and children. The graves are shallow and unlined; each included from one to many individuals. Within each tomb, the bodies

**ABOVE** Many Chinchorro mummies, like the one shown here, were of children. The litters they were carried on are as well preserved as the mummies themselves.

**BELOW** Chinchorro mummies as much as 7,000 years old retain the facial features modeled in clay by the relatives of the deceased.

ABOVE Radiographs and other noninvasive analysis of Chinchorro mummies have helped researchers understand how the mummies were made.

are placed in an extended position and wrapped, often limb by limb; they are most often laid on the back but sometimes on the side, and lower limbs are sometimes slightly flexed. Grave goods are relatively sparse and can include textiles, skins, bags and implements of bone, stone and other materials. Many of the items are recognizable as fishing and shell-fishing gear, including fish hooks, nets and net weights, and prizing tools for shell-fishing.

Studies of the mummies reveal a great deal about how the Chinchorro people lived, their general state of health and other details of their lives. Based on the grave goods alone, many early researchers argued that the Chinchorro were primarily fishers and coastal (intertidal and subtidal) gatherers. A common pathology visible in the skulls of the Chinchorro, especially males, is a growth in the ear known as *external auditory exostoses*, which is thought to be caused largely by continual diving in cold water. This finding supports the idea that diving for fish and shellfish was a major activity. Dietary studies based on analysis of trace elements in bone and hair, and studies of coprolites (feces), however, indicate that the Chinchorro had a mixed diet, consuming plants and animals from both the land and the sea.

The overall health of the Chinchorro population appears to have been relatively good for an ancient hunting and gathering society. The mummies show some incidence of bone lesions caused by nutritional stress and possible iron deficiencies but, in general, Chinchorro remains show little evidence of either illness or violence. One interesting note is that some of the skulls of the Chinchorro, especially the later ones, show evidence of having been artificially deformed by wrapping. The practice of artificial cranial deformation was widespread in the Andes by the time of the Inca. Apparently its origins lay thousands of years before among the Chinchorro, and perhaps other early people.

RIGHT Chinchorro mummies are often found in groups, as is shown in this museum display reconstruction of a Chinchorro tomb. The range of age and sexes represented has led some researchers to speculate that the Chinchorro were buried in family groups.

Studies of Chinchorro mummies have also revealed how they were made, and researchers studying the mummies have assigned them to categories based on how they were prepared and what they looked like. The first category is the natural mummies – that is, those that formed naturally from human bodies that were not prepared in any way. These mummies were preserved naturally in the dry climate of the Atacama Desert. The second category comprises those prepared in the most complex manner, including evisceration and reconstruction. The third category are mummies formed by encasing unprepared bodies with a layer of clay. Mummies of both the second and third categories were often painted with red or black pigment.

One of the most interesting features of Chinchorro mummy practice is that many of the most elaborate were of children and babies. In addition to actual mummies prepared from whole bodies of children, the Chinchorro also made small figurines in the form of mummies. These figurines sometimes contained bones from infants or fetuses, or even animals. It may be that the small bodies of babies were too difficult to mummify using the normal techniques and that mummification of the smallest of the Chinchorro required special techniques.

Some researchers have argued that the different types of mummies were made at different times, and that practice gradually progressed from simple to complex. In many cases, however, mummies of several categories are found buried in the same tomb. This means that either the different kinds of mummies were being made at the same time or that tombs were opened and reused over hundreds and even thousands of years, and perhaps that the mummies were pulled out of the ground, or kept above ground, for many years. It is true that some of the mummies show signs of having been repaired and repainted, so perhaps they were kept and used in rituals by the Chinchorro.

**ABOVE** The surfaces of Chinchorro mummies were generally covered in pigment or paint. This is one of the more unusual mummies, with painted stripes still visible. It is possible that the original surfaces of more mummies were decorated in a similar way, but that they have deteriorated over the thousands of years since they were made.

**BELOW** The preservation of hair on the mummies has allowed researchers to study the evolution of pre-Columbian hairstyles in the Andean area.

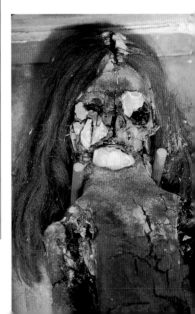

The origins of Chinchorro mummification practices have been subject to scholarly debate for some time. Some researchers have argued that the practice must have come from the jungle. Recent excavations in northern Chile, however, have demonstrated that the Chinchorro practice of mummification has its roots in the burial of wrapped and stiffened bodies nearly 9,000 years ago. This indicates that the artificial mummification practices of the Chinchorro developed locally and were invented by the small-scale hunter-gatherer-fishers who lived on the coast of the south-central Andean region thousands of years ago, before pottery, agriculture and even settled village life.

How did small groups of hunting-gathering-fishing peoples develop such a sophisticated and specialized set of mortuary practices? It is probable that they got the idea from the observation that the arid climate in which they lived preserved bodies almost indefinitely, as long as the remains were protected from scavengers. Hunting peoples generally develop a quite sophisticated knowledge of anatomy, and it is not difficult to imagine that the Chinchorro simply began applying this knowledge to the making of mummies. Understanding why they bothered to do so is less easy.

While we may never know for certain which is correct, several interesting theories have been developed about why the Chinchorro made mummies. One theory holds that the dead had an important role in this small-scale, family-based society. The suggestion is that mummified remains were carried with these semisedentary people as they moved from place to place, and that the mummies were placed in positions of honor and importance during major rituals. Another possibility is that mummies were kept until the group moved back to their coastal base camp, or until enough members of a single family had died to make the burial ritual worthwhile. This theory is consistent with the large number of highly treated mummies of children; children probably died in higher numbers than adults, but they may not have been eligible for burial until other members of their families died.

The development of complex artificial mummification in a small-scale society remains an extraordinary archaeological event. But perhaps it was not such an extraordinary historic event. The coast of northern Chile provides some of the best conditions for archaeological preservation anywhere, and items buried in tombs even thousands of years ago appear to have been made yesterday when they are excavated. In nearby areas of southern Peru, with preservation conditions only marginally poorer, the traces of Chinchorro mummies are faint. The remains of clay masks and figurines are found, along with skeletal remains of burials that fit the pattern of Chinchorro tombs, but few well-preserved mummies exist. This leads us to the question of whether many ancient hunting and gathering communities may have had extraordinarily complex mortuary rituals that are simply invisible in the archaeological record because most areas lack the conditions to preserve the evidence so perfectly.

**BELOW** One of the mummies used in the most important early study of Chinchorro mummies, by the German researcher Max Uhle. His research on this mummy, which was part of a collection he made of about 100 mummies, was first published in 1919.

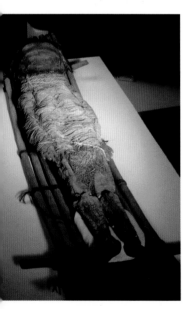

# The Mysterious Mummy in
## Tomb 55

O n January 6, 1907, an archaeological mission, led by British Egyptologist Edward Ayrton and financed by the American lawyer and amateur archaeologist Theodore Davis, discovered the modest entrance to tomb KV (King's Valley) 55. In plan, the tomb was simple: A sloping corridor led downward to a single room that still housed a mummy in a battered inlaid coffin. The tomb's contents, however, were far from simple, and the bare walls offered no clue as to its occupant. It baffled its discoverers and continues to be a topic of hot discussion among Egyptologists today. Just who was buried in KV 55?

## The Tomb

The burial chamber was a mess, with boxes, mud bricks, builders' debris and workmen's tools scattered carelessly around. In the middle of this chaos was a golden inlaid coffin whose wooden bier had collapsed, tipping the coffin and its contents onto the floor. Beneath the dislodged wooden lid the archaeologists could see a mummy, while a niche in the right-hand wall held four human-headed canopic jars. Everything was frosted with glittering motes of gold that had floated from the decaying remains of a dismantled gilded wooden shrine propped against the walls.

The tomb was anonymous and contained little written material, although its objects were variously labeled with the names of the late eighteenth-Dynasty pharaohs Amenhotep II, Amenhotep III, Queen Tiy, Akhenaten and Tutankhamen. As Tutankhamen's was the last name to feature, it seemed that the tomb must have been closed during or soon after his reign.

The late Eighteenth Dynasty was a confused time. The heretic pharaoh Akhenaten had stunned his people by rejecting the traditional gods, abandoning Thebes and retreating with his court to a new capital city, Akhetaten, today better known as Amarna. Here he spent his days in splendid isolation, worshipping the sun god, Aten. He and his consort, Nefertiti, had six daughters but, as far as we know, no son. After 17 years on the throne, Akhenaten was succeeded by the young Tutankhamen. Almost immediately, the new king abandoned Amarna, reinstated the old gods, and returned to Thebes. Tutankhamen never specified his parentage, but as he succeeded to the throne as a child, it seems safe to assume that he was of royal blood, almost certainly the son of Akhenaten and his favorite secondary wife, Kiya.

**ABOVE** Amenhotep IV, better known as the hertic pharaoh Akhenaten, had a distinctive appearance. His statues reveal his long face and chin, narrow eyes, sensuous lips, meager shoulders, spindly arms, and wide hips.

Akhenaten's family had all been interred in the royal tombs cut into the Amarna cliffs. But, as grave goods gathered from several Amarna tombs are included in KV 55, their burials must have been reopened and at least some of their contents transferred to Thebes during the reign of Tutankhamen. KV 55 has yielded a set of magical funerary bricks inscribed for Akhenaten's funeral, while the dismantled gilded shrine proved part of the burial equipment provided for Queen Tiy by her son. The four canopic jars bore ill-fitting, and therefore possibly not original, stoppers whose delicate female heads sported the bobbed wigs worn by Akhenaten's daughters and by Kiya.

## The Coffin

The anthropoid coffin had initially been built for a nonroyal woman who wore a wig rather than a crown, perhaps Kiya. It had subsequently been remodeled and fitted with a beard and uraeus (cobra), making it suitable for the burial of a royal male. The uraeus and the gold mask that covered the face had then been torn off in antiquity, leaving just one eye in place and much of the underlying wood exposed. The text on the coffin-foot and the bands of hieroglyphic decoration were originally written as prayers to be spoken by a woman. However, these inscriptions had been altered from feminine to masculine, while the name of the original owner had been replaced by a name in a cartouche or royal ring, which was itself subsequently erased.

## The Mummy

Nothing about the mummy proved that this was a member of the royal family, but its burial in the Valley of the Kings and its prestigious grave goods strongly suggested that this was an Amarna royal. Under Davis's supervision, the mummy was hastily stripped of its wrappings, but to everyone's disappointment, it was discovered to have been damaged beyond repair. For many centuries, water had been dripping via a crack in the ceiling directly into the open coffin, and the mummy's flesh had completely rotted away. The skull, which had been further damaged by a fall of rocks, was subsequently reconstructed.

No photographs were taken of the unwrapping, and we have to rely on eyewitness accounts to reconstruct the image of a small body, wrapped with the left arm bent with the hand on the breast and the right arm straight; this, halfway between the crossed arms of the pharaoh and the straight arms of the commoner, is a standard royal female pose. The mummy wore a pectoral and had inscribed gold bands within its bandages, but these were stolen before they could be properly recorded. The mummy, now a skeleton stained brown by embalming resins, was sent to Cairo Museum, where it remains today.

## Tiy, Akhenaten and Smenkhkare

Theodore Davis was convinced that he had discovered the body of Queen Tiy, mother of Akhenaten and wife of Amenhotep III. Two "experts," a local doctor, Dr. Pollock, and an anonymous American obstetrician enjoying a vacation in Luxor, agreed with him that the body was female, and so it was as the "Tomb of Queen Tiye" that Davis published his discovery in 1910. He never changed his opinion.

But in 1912, the anatomist Grafton Elliot Smith emphatically announced that the skeleton was male. Every expert who has since examined the bones has agreed with this reassessment. Arthur Weigall, chief inspector of antiquities at Luxor at the time the tomb was discovered, put forward the intriguing suggestion that the bones belonged to Akhenaten himself. Smith happily agreed and was prepared to revise his own appraisal of the skeleton's age at death, originally given as 25 years or less, in order to fit Weigall's theory. However, this interpretation of events raises several important questions. Why, for example, was Akhenaten buried in Kiya's secondhand coffin? Why was Tiy's shrine included in her son's tomb? How old was the body in KV 55?

Akhenaten had reigned for 17 years and had fathered a daughter during the earliest years of his rule. He therefore could not have been less than 30 years old at his death and was likely to have been considerably older. If this skeleton were really Akhenaten, it had to be that of a middle-aged man. But when, in the 1920s, the anatomist Douglas Derry restored the bones, he classified them as the remains of a young man whose unfused epiphyses and an unerupted right upper third molar showed that their owner could have been no more than 25 years old. This was the age originally accepted by Smith.

In 1963, Professor Robert Harrison of Liverpool University reexamined the skeleton and drew the same conclusion. The remains were those of a young male who shared the same blood group and the same slightly elongated head shape as Tutankhamen. This similarity of skull shape has recently been confirmed by Dr. James Harris who, after studying computerized tracings, suggested that Tutankhamen and the KV 55 body were first-degree relatives, either a father and son or full brothers. Tutankhamen never fathered a son. The body must therefore be that of his father, almost certainly Akhenaten, or his brother, the ephemeral Smenkhkare, a prince who made a fleeting appearance to marry Princess Meritaten and rule as coregent alongside Akhenaten, and who had disappeared just as Tutankhamen came to the throne.

BELOW The mummy had been badly damaged by water dripping into the coffin through a crack in the tomb roof. Today it is reduced to a disarticulated collection of bones.

ABOVE Dr. Joyce Filer and Dr. Nasri Iskander prepare to re-examine the KV 55 bones, in an attempt to determine once and for all their gender and their age at death.

BELOW Buried alongside the mummy was a confusing jumble of grave goods taken from other burials. The canopic jars, which should have held the vital organs of the deceased, were provided with stoppers originally carved for a woman.

## Examining the Bones

The KV 55 bones hold the key to understanding the complexities and confusions that cloud our understanding of the end of the Amarna age and the start of Tutankhamen's reign. The most recent analysis on the KV 55 bones was conducted in the year 2000 by leading physical anthropologist and Egyptologist Joyce Filer. She confirmed that the skeleton is male and ascertained that the bones, although now stored in separate boxes, do indeed belong to the same body. Having examined both bones and teeth and studied the available X-rays, she concluded that the remains are those of a younger man who died in his mid-twenties or younger and who, while living, bore a strong resemblance to Tutankhamen. It was not possible to suggest a cause of death, but the skeleton shows no obvious sign of abnormality.

The consensus of expert opinion is therefore clear. The bones are those of a young man, almost certainly those of the lost elder brother of Tutankhamen, Smenkhkare.

## Reconstructing the Past

We may now attempt a reconstruction based on the evidence of the bones. Tutankhamen now lived in Thebes, where he worshiped the traditional gods. He was prepared to abandon Amarna and all that it stood for. However, he was not prepared to abandon his dead brother, who was still buried in the Amarna cemetery. Troops were sent back to Amarna to retrieve what they could. The royal graves had already been thoroughly looted, but Smenkhkare's mummy was saved, as were a variety of grave goods. The better artifacts were set aside for use in Tutankhamen's own tomb; one of his three coffins and his canopic jars originally belonged to his brother Smenkhkare. The less valuable goods were put together in a semblance of a royal burial for Smenkhkare, who was secretly reinterred in the appropriate safety of the Valley of the Kings. Smenkhkare was, however, irretrievably smeared with the tarnish of Amarna heresies. At the last moment, his name was defaced in his tomb by the necropolis priests, who were determined to erase the last evidence of the hated Amarna period.

# Restoring the Royal Mummies

B y the Twenty-first Dynasty (1070–945 BC), security within Egypt's cemeteries had broken down. The New Kingdom royal tombs had been looted, and the pharaohs lay desecrated and vulnerable. The Third Intermediate Period priests of Amen decided on a rescue mission. It is thanks to their work that the New Kingdom mummies have survived.

## Tomb Robbers

Tomb robbers, the scourge of the cemeteries, ensured that few lay undisturbed. All too often, the robbers were the gravediggers, whose detailed knowledge enabled them to target the richest pickings. Even the undertakers were not above suspicion; several mummies were robbed before their tombs were sealed. The penalties for those caught stealing were severe. Suspects were interrogated with torture, and the guilty faced a lingering death impaled on a stake. But at times of high inflation and civil unrest, this did little to deter the determined criminals, and the situation went from bad to worse.

## The Royal Rescue

The priests were concerned. Not only were the robbers desecrating the dead, they were stealing state property. A rescue plan was devised, and the mummies were made immune to attack. They were stripped of their gold and their tombs emptied.

Gathering the tattered remains of the pharaohs, the priests transferred them to workshops in and around the Valley of the Kings. Here they repaired the mummies as best they could, reattaching limbs before rewrapping them in contemporary bandages and replacing them in their wooden coffins, now stripped of all gold leaf. Both bandages and coffins were labeled meticulously, and all groups of mummies were put into storage. Eventually there were two main collections, one in the Pinodjem II family tomb at Deir el-Bahari and the other in the Valley tomb of Amenhotep II.

## Discovery at Deir el-Bahari

In 1871, Ahmed el-Rassul was searching for a lost goat when he stumbled across the Pinodjem tomb. Sensing an unparalleled business opportunity, Ahmed dropped a dead donkey down the shaft. The ripe smell of decay was meant to keep Egyptologists away.

The New Kingdom mummies had nothing left to offer. Ahmed and his brothers, notorious tomb robbers, bypassed the impoverished pharaohs and targeted the Pinodjem family, stealing and selling a series of illustrated papyri and funerary figurines. Egyptologists quickly realized that a new tomb had been discovered, but it was 10 years before an investigation was launched.

The el-Rassul brothers were the obvious suspects. Eventually Mohammed, the eldest brother, decided that the time had come to confess. On July 6, 1881, Egyptologist Emile Brugsch was lowered into the Pinodjem tomb. Here was an amazing sight: Egypt's greatest kings, including Ahmose, Tuthmosis I, II and III, Ramses I and II and Seti I. Brugsch decided that the bodies should be transported at once to Cairo. No plans were drawn, and no photographs taken. Instead, the mummies were taken from the cool dark tomb, their home for almost 3,000 years, and winched into the hot sunlight. Soon each pharaoh, wrapped in protective matting and sewn into sailcloth, was sailing northward. Crowds gathered along the Nile to watch the cortege, the women weeping and tearing their hair in the traditional gesture of mourning.

When, in 1898, Victor Loret discovered the lost tomb of Amenhotep II, the collection of New Kingdom pharaohs was almost complete. The one obvious omission was Tutankhamen, who was discovered by Howard Carter in 1922 (see p. 103). An open sarcophagus yielded the mummy of Amenhotep II, a side chamber held three mummies and a large side room supplied nine named coffins. Here were Tuthmosis IV, Amenhotep III, Seti II, Siptah and Ramses IV, V and VI.

## Unwrapping the Bandages

Both bandages and coffins had been labeled in hieratic script, on the basis of which we are able to put names to the otherwise anonymous kings and queens. Unfortunately, in some instances, the mummies and their coffins had been mixed up. Ramses IX, a Twentieth-Dynasty king, lay in the coffin of the Third Intermediate Period Lady Neskhons. The coffin of the Eighteenth-Dynasty Queen Ahmose Nefertari also included

**BELOW** The mummified kings and queens of the New Kingdom were robbed of their valuables within decades of their deaths. A rescue mission, conducted by the cemetery priests, ensured that they were re-wrapped and given a secure burial.

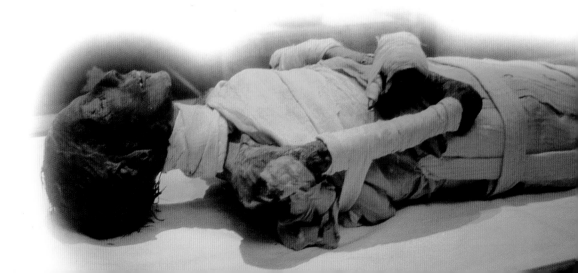

the Twentieth-Dynasty Ramses III, while the Eighteenth-Dynasty
Ahmose-Inhapy lay in the coffin of the Eighteenth-Dynasty nurse Rai.

The mummies, then, were occasionally stored in the wrong coffins,
but we have little reason to doubt the vast majority of the labels scrawled
on their bandages. "Ramses II," known to have died in his nineties, was
indeed a very old man. "Sekenenre Tao II," who died defending his land
against the Hyksos, has horrific head wounds and the characteristic
marks of a Hyksos battle axe imprinted on his skull. In just a few cases is
identity in doubt. The badly damaged mummy attributed to Amenhotep
III is thought by many experts to have been misidentified, while the
mummy attributed to Tuthmosis I is likely, on the basis of the positioning
of the arms, to be the remains of his nonroyal father, Ahmose Sipairi.

## Tuthmosis III

Tuthmosis III was the first of the Deir el-Bahari pharaohs to be
unwrapped. This was by no means his first exposure. Tuthmosis had been
attacked by New Kingdom thieves and completely stripped and rewrapped
by Third Intermediate Period restorers. His bandages had been slashed by
the Abd el-Rassuls and, in 1881, Brugsch had unwrapped and rewrapped
him, discovering the four wooden supports incorporated in his bandages
as stiffening. Now, in 1886, Gaston Maspero, director of the Egyptian
Antiquities Service, took his turn. Slicing through the bandages in front
of an invited audience, he revealed the face of "the Egyptian Napoleon."

Tuthmosis was in pieces. His hands were folded across his chest,
but his head and limbs were detached and his penis and testicles were
missing. He had died an elderly, bald man of medium build standing some
5 feet, 5 inches (165 centimeters) tall. His head was large, but his well-
preserved face was small and narrow, equipped with a high-bridged nose,
a low forehead and buck teeth that showed some wear but no decay.

Thirty years later, Tuthmosis was reexamined by the anatomist
Grafton Elliot Smith. He was then left in peace until 1968, when he took
part in a research program devised by the University of Michigan School
of Dentistry under Dr. James Harris. The team used a combination of
X-rays and cephalometrics – radiographs that, taken with a precise
orientation, facilitate comparisons between mummies. The investigators
were able to identify a strong family resemblance between Tuthmosis III,
his father Tuthmosis II, his grandfather Tuthmosis I (or, more accurately,
his great-grandfather Ahmose-Sipairi) and his son Amenhotep II.

Attempts to use X-ray analysis to determine Tuthmosis's age at death
were, however, a failure. His X-ray age of 35 to 40 years is incompatible
both with the historical records, which give him a reign of at least 50
years, and with his physical remains, which are those of an elderly man.
It seems that the technique is unreliable, and this is not the only instance
where X-ray analysis has suggested an unacceptably early age of death
for a mummy. The most striking example is the dating of the Nineteenth-

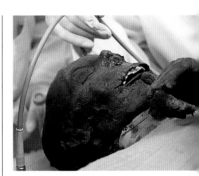

**ABOVE** The collection of royal
mummies held in Cairo Musuem
offers archaeologists an invaluable
bone and tissue bank. Using this
material it has been possible to
prove genetic links between
members of the 18th-Dynasty
royal family.

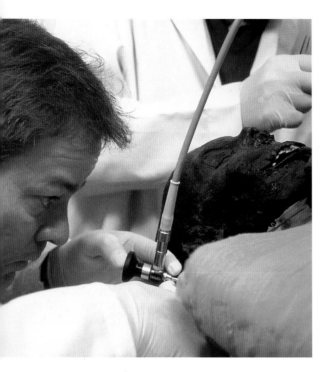

Dynasty Ramses II, who had a well-documented reign of 66 years and an extremely old body, and yet who apparently died aged 50 to 55!

The most recent investigation into Tuthmosis III was conducted by a team from Utah, working with Nasry Iskander of Cairo Museum. Scott Woodward and Wilfred Griggs developed their expertise in the extraction of ancient DNA working in the Greco-Roman Faiyum cemetery. They have taken tissue samples from Tuthmosis's body cavities – which, it is hoped, have not been in contact with modern DNA. Their work is still in progress, but already it has confirmed some of the long-known historical facts about the eighteenth Dynasty. The DNA profile makes the dynastic break between Amenhotep I and Tuthmosis I, his adopted son, obvious.

Tuthmosis has now been sealed in a nitrogen-rich display case, and will not be available for further study in the foreseeable future.

**ABOVE** Archaeologist Scott Woodward extracts a tissue sample from the mummy of Tuthmosis III. He is using a noninvasive technique which will not damage the already battered body further.

## Body in Sheepskin: An Ancient Murder Mystery?

Maspero quickly cut his way through the rest of the royal mummies, with varying results. Ramses II was in good condition, but Ramses III was coated in a thick black resin that had to be chipped away with scissors. Something terrible had happened to Queen Ahmose Nefertari, who was slimy, and stank. She had to be hurried from the operating table and was eventually buried beneath the museum storehouse until the smell died.

The Third Intermediate Period priests had treated the ancient pharaohs with respect, and Egypt's kings and queens lay calm and dignified beneath their bandages. One body, however, was anything but serene. A rolled sheepskin discovered in a plain wooden coffin at Deir el-Bahari yielded an anonymous male body, bound hand and foot, whose face was twisted in a grimace of unspeakable anguish. The unnamed man had not been eviscerated, and all his internal organs were still in place. Traces of dry natron found on the skin suggest a sketchy attempt at artificial embalming, but it was the dry sheepskin rather than the desiccating natron salt that had drained the fluids from his body, allowing natural mummification to occur.

As this body was found in a royal cache, we must assume that it is that of an important male. Royal males, however, were not wrapped in sheepskin, which was considered unclean and seldom worn even by the living. His face suggests that he met with a horrible death. Perhaps he had been a traitor, even the prince whose failed coup caused the assassination of Ramses III and who is said to have committed suicide.

# Funerary Rituals of the
## Tashtyk Culture

I n 1772, Peter Simeon Pallas, a natural history professor, botanist and ethnographer, and a participant in the scientific expedition sent by Catherine the Great to Siberia, described various ancient sites in the steppes of the Yenisei. He listened to everything he was told by grave-robbers (whose "profession" was at that time widespread in Siberia), taking down even what seemed to be fantastic tales. One grave-robber told him that he had seen in graves "life-size human heads, empty inside, made of a porcelain-like paste and painted with red and green leaves." Amid many legends connected with kurgans (Siberian burial mounds), rumors about such "living faces" were circulating among peoples. A hundred years later, one such head, an unpainted specimen, was brought to the Museum of Minusinsk by a peasant who found it in a destroyed vault on a riverbank. These same graves were excavated at last by a professional archaeologist, Alexander Adrianov, in 1883, and immediately became a sensation. In a big vault, he found about 33 "heads" – masks on the skulls of skeletons – and busts next to human ashes. Later, he also excavated some small timber structures that contained one or two burials each, sometimes with the remains of a cremation, but with masks as well. However, the main discovery was to be made in 1903.

ABOVE What the locals referred to as "living faces" turned out to be funerary masks of the Tashtyk culture dating back as far as the sixth century AD.

## Early Discoveries

On the left bank of the great Siberian river, the Yenisei, not far from the town of Abakan, is a mountainous massif called Oglakhty. It is famous for the presence of numerous archaeological sites from different periods; it also has the biggest complex of rock art in southern Siberia. Here, incredibly well preserved graves of the Tashtyk culture were discovered. This mountainous area is criss-crossed by numerous gullies. On the slope of one such gully, in 1902, a shepherd was looking for a

LEFT Many of the funerary masks were finely polished and painted decoratively, often in red.

ABOVE The masks were made in the likeness of the deceased and had been placed over their faces.

lost foal when suddenly something was crushed under his horse's legs. The horse jumped aside, and the shepherd went down into a hole. Looking around, he found himself in an ancient vault. Two corpses were staring at him, one baring its teeth, the other with a painted mask on its face.

The horrified man ran away to his village and told his mother-in-law, who was known to be a brave and practical woman. She knew about treasures from ancient graves (looting them was a profitable business in Siberia) and rushed back to the place. She ascertained that the face of one corpse was covered with green silk under the mask, and that a nearby wooden bench had a suede purse and broken pottery on it. She found a mannequin filled with dry grass on the floor; it was wearing a suede jacket. She did not find golden objects, but she took the purse and the mask and managed to sell them later.

Rumors about "the people buried alive" in the kurgan soon reached Adrianov. He received permission and an assignment from the Imperial Archaeological Commission and, in 1903, started excavating. He found two other tombs in an ideal state of preservation. The timbers of the floor and ceiling were made of big larch logs, so durable that the workers could saw and chop them up for wood. Apart from ceramic vessels, he found wooden plates, cups, bowls, scoops, kegs and buckets. Smaller wooden objects had also been preserved, such as models of bows, daggers and swords, even hairpins. He also found remains of clothes made of fur and leather: hats, coats, belts and ribbons. The remnants of mummified bodies, human ashes, bones and skulls were lying on the floor with "pillows" under their head – stones, pieces of log or leather pillows filled with grass. All the skulls were covered with funeral masks, finely polished, fully painted in red, or decorated with spirals.

Other valuable finds were life-size funeral dolls dressed in the same way as the real dead – mannequins whose torsos, legs and arms were made of leather bags filled with grass. The heads were made of grass lumps covered with leather and silk, with modeled facial features. Two of them had plaster masks. The dolls had fallen to pieces and gotten mixed up with the skulls and ashes. A hairpiece and a plait had also been preserved, but it was hard to establish whether they belonged to the humans or the dolls.

During the twentieth century, Russian archaeologists excavated hundreds of burials of this type in southern Siberia. It was established that the Tashtyk culture, named after a river in Khakassia, existed in this forest-steppe area in the first to sixth centuries AD – a period of a great economic, social and especially ethnic change. The complexity of the ethnogenetic processes and a mixture of cultural traditions are reflected in the Tashtyk culture's unusually complicated funeral practices.

## Pit Graves and Burial Vaults

The Tashtyk people built two very different types of funerary structure: cemeteries of pit graves and burial vaults. The cemeteries, with dozens or even hundreds of graves, were placed on the slopes of hills, valleys or gullies, or occasionally on the high mounds of earlier kurgans. The vaults, in groups of two to four, are sometimes located near the grave cemeteries, but they are always clearly isolated and higher up the slopes.

The pit grave cemeteries are barely visible. All were found by chance, as they had minimal constructions on the surface, unlike the other burial sites of the Yenisei steppes. The Tashtyk people dug the graves in the form of rectangular holes 5 to 10 feet (1.5 to 3 meters) deep. They covered the walls and floor with a large expanse of specially prepared birchbark. Then they built a low timber structure and, after inserting the deceased (usually one or two corpses, and never more than four), closed it with logs, birchbark and soil.

The funeral vaults, on the other hand, are clearly visible on the surface due to their square mounds of stones and earth. Thanks to well-developed methods for excavating them, it was possible to reconstruct the sequence of their construction in every detail. First, a big rectangular ditch, 3 feet, 3 inches (1 meter) deep, was dug, and a timber structure of massive logs, with a floor and a ceiling, was built inside. The timber walls were consolidated with stone slabs, and the entire construction was surrounded by stone slabs as well. The whole thing was covered with birchbark, and the surface mound took the shape of a truncated pyramid. Special entrances, usually from the west, in the form of stairs or an incline, were consolidated with vertical logs or big stone slabs. The size of the chambers varied from 52½ to almost 300 square feet (5 to 30 square meters) and ranged from small (for 10 to 40 people) to very large (for 100 people or more). Presumably, the first corpses were seated along the walls, sometimes on special planking, and then the central part was gradually filled in.

After the vault had been filled with bodies, funeral dolls, separate skulls and, mostly, cremations, it was set alight, presumably to let the souls of the deceased ascend to the "land of their forebears" with the flames. A big fire was set on the roof, which sometimes crashed down, burning out all the wooden constructions, but sometimes the fallen roof simply blocked the chamber, which produced a very high temperature while precluding oxygen access. The smoldering and carbonization allowed many organic objects to be preserved, such as wooden and birchbark utensils, bags, wooden sculptures of animals and humans, pieces of vehicles and unique planks with finely engraved battle and hunting scenes.

**BELOW** Straw-filled mannequins were discovered among the mummies. On closer inspection, the most complete examples were found to contain a bag of human ashes in the chest area.

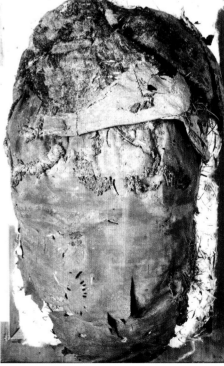

## Social Status

The diverse Tashtyk funerary rituals were different not only for adults and children but also for various categories of adult. The pit graves belonged to family groups and the vaults to clans or dynastic groups. The vaults were intended for adults only; child burials were placed separately in small timber structures nearby. An interesting arrangement was discovered recently in a vault. In the middle of the timber structure was a wooden platform for sacrifices, with the remains of more than 30 cremated bodies around it; to the left and right of the entrance were pits containing odd human skeletons, probably remnants of sacrifices. Accompanying burials were found in other vaults, probably also sacrificial in nature. This makes it possible to propose that the vaults were constructed for people who had a higher social status than those in the pit graves.

Owing to the conditions of the Siberian climate, burials could be carried out in the warm season only – from spring, when the ground thawed, to autumn. Tashtyk funeral practice embraced several modes of burial: the so-called primary burials of a corpse or mummy; secondary burials, when mummies or skeletons were reburied; and the cremation of the deceased and burial of its ashes. These funerals involved long preparations, especially when it became the rule to bury two or more dead people at the same time in the same timber structure. In the early stage of the culture, bodies were most often buried, but gradually cremation became more widespread.

To preserve a body until it could be buried, the Tashtyk people used embalming or mummification. We know only their methods of preserving heads: (1) covering them with gypsum, (2) extracting the brain (as is known from the presence of post-mortem trepanation of the skulls), or (3) a combination of these methods. The trepanation was carried out by specialists. Adrianov noted that pieces of skull found at Oglakhty were bashed in with "an adroit and practised stroke." For thin-walled skulls, pieces measuring 1 to 1.4 inches (2.5 to 3.5 centimeters) were bashed in. For thick-walled skulls, the preservers drilled several little holes, then gouged out (or pressed in) the piece between the holes with a sort of chisel. Anthropologists who examined the skulls from one cemetery proposed that the trepanations had all been executed by the

**BELOW** These images show that post-mortem trepanation was carried out on the skull in order to remove the brain prior to mummification.

same person. The bone was struck out either in the occipital or the sincipital area. Perhaps this depended on the surgeon, as different positions of trepanation predominate in different cemeteries. The piece of bone is often inside the skull.

Bodies were probably embalmed without serious surgical operations, and only the internal organs taken out. At Oglakhty, parts of three naturally preserved mummies were found in 1903. The male mummy had part of its facial skin, an intact neck and remains of skin, muscles and blood vessels on the vertebrae. The female mummy had facial skin, one breast with a nipple, a shoulder and both arms. The skin showed no traces of cutting or sewing. The male mummy had his left eyeball, but sometimes the eyes were taken out or pressed in, and the sockets filled with gypsum.

Cremation was a very different process. The Tashtyk rite, which involved cremating bodies and burying the ashes in various packages, was a new funerary practice for the Siberian tribes, possibly connected with groups recently arrived in the region. The cremation was rather crude; each body yielded 2.2 to 4.4 pounds (1 to 2 kilograms) of ashes, with pieces of bones up to 4 inches (10 centimeters) in length. This has enabled anthropologists to determine that only adults were cremated, and in every case where it was possible to sex them, they were males.

When the first mannequins were found in pieces at Oglakhty, they were not associated with the deceased. For a long time, they were considered to represent wives or servants accompanying the dead to the other world, a widespread custom in Chinese funerary practice. It was only after two complete mannequins were found in 1969 by Leonid Kyzlasov in the same cemetery at Oglakhty that their meaning became obvious. X-rays revealed a little bag of human ashes in the chest of each mannequin. Thus the mannequins, or funerary dolls, imitated the body of the cremated person himself and served as a receptacle for his ashes, a sort of a funerary urn. The Tashtykians used leather or sheepskin bags filled with grass to make the dolls, sewing the parts together into the shape of a human body; sometimes they simply sewed together a jacket with trousers, likewise filled with grass. Then they dressed the dolls in the clothes of the deceased, just as they did mummies, and attached the original plait to the head, or put the scalp on it.

Although few dolls have survived, their original presence in most of the vaults is easy to deduce. The place, position and orientation of those buried were highly regulated; the bodies, mummies or dolls were laid down with their upper portion raised, or were seated with legs

**ABOVE** Human hairpieces and plaits were found among the mummies and mannequins, although archaeologists remain unsure to which they belong.

ABOVE Various grave goods were discovered in the tombs, among them ceramic vessels (as here), wooden plates, cups and bowls.

outstretched and parallel, facing east. When the leather and grass of the dolls decayed, the pile of ashes still lay parallel to the chest of the nearby skeleton. The position of vessels and of bones from pieces of meat also proves the presence of dolls.

Both mummies and dolls were dressed in winter clothes, with artificial decoration. Presumably they were kept, completely dressed, for some time in something like a shrine or house of the dead. At any rate, the mummies were sometimes buried in a dilapidated state and with mended masks. Although everyday clothes were used (there is a patch on the sleeve of one fur coat from Oglakhty), they were turned into ceremonial funerary robes by the use of wooden decorations covered with gold leaf, plus lots of glass beads and gems. Other grave goods included wooden models of weapons (daggers and knives in scabbards, bows, quivers), horse harnesses, mirrors and small cauldrons – also wooden and birchbark crockery and ceramic vessels. The most common finds are bone hairpins for the complicated Tashtyk coiffures.

## The Funerary Mask

But the distinguishing feature of the Tashtyk funerary ritual, character-istic both for pit graves and vaults, was the funerary mask. The masks were not actual death masks, cast from the face of the deceased. They were sculpted onto the faces (some have imprints of wrinkles and hair on the inside) or on special soft block-heads. Most often, the face was covered with a piece of silk, and then thin layers of gypsum were laid onto it, one after another, modeling the face. Some also had the neck and ears modeled, and some were busts, which stood in burial chambers on special stands. The masks are thought to be faithful, individualized likenesses of the people whose faces they covered. Made of a mixture of porcelain clay, gypsum, limestone powder and quartz sand, they were painted red and black. Women's masks had red spirals or other patterns on the forehead, temples, cheeks and chin, while males had black patterns on a solid red background. These patterns may perhaps represent tattoos that the deceased bore in life.

One needs to note that this tradition changed over the course of the period. Toward the end, the busts became more popular and widespread, and their decoration became more complicated. Green and light blue paint was also used; eyelashes, bead necklaces and pendants were modeled or painted. Recent discoveries have revealed human ashes inside some busts, so they were probably also used as funerary urns, possibly replacing the dolls in this capacity at the end of the period.

Why were the funerary rites of the Tashtyk culture so complicated? Why is there such great diversity in the methods of burial, of funerary constructions, and so on? Chronological, social and ethnic causes have all been cited; most likely there was a combination of such factors. Early Tashtyk cemeteries contained more or less equal numbers of bodies and cremations but, in the later ones, cremations are clearly dominant. Among the bodies, most are women, but among the cremated, women are never found. The anthropological type of the male bodies did not differ from that of the women – they are Europeoids with a Mongoloid admixture. So it seems logical to propose that cremation was applied to men only and, moreover, to men who were not of local origin. Masks from the graves represent two anthropological types, Europeoids and Mongoloids, while masks from the vaults present a third type, with a mixture of features, as well. This demonstrates not only the spread of the Mongoloid type but also the formation, because of that, of a new anthropological type of population. One way or another, the Tashtyk funeral sites reflect differences in the social status of people, but the criterion of social distinction is hard to determine. Was it wealth, noble birth, or membership in an ethnic group that dominated in this society?

**LEFT** Several of the masks, and those of men in particular, revealed black patterns over a solid red background. It is thought that these represent tattoos.

**BELOW** This image clearly shows how much of the facial skin remains intact, and traces of muscle and blood vessels can be seen in the neck area.

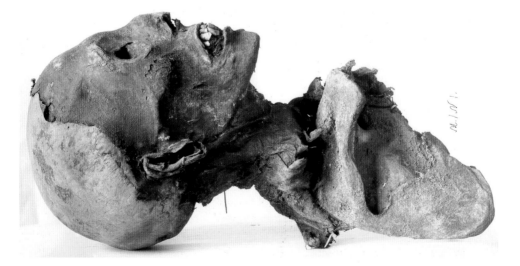

# Index

Page numbers in *italic* type refer to picture captions; **bold** page numbers refer to major entries, which will include both text and illustrations.

# Bibliography

**The Mohenjo Daro "Massacre" (pp 10–14)**

Dales, G. "The Mythical Massacre at Mohenjo-Daro," *Expedition* 6 (3, 1964): 36–43.

Kennedy, K. A. R. "Skulls, Aryans, and Flowing Drains: The Interface of Archaeology and Skeletal Biology in the Study of the Harappan Civilization." In *Harappan Civilization*, edited by G. Possehl. New Delhi: Oxford and IBH Publishing, 1982.

Kennedy, K. A. R. *God-Apes and Fossil Men: Paleoanthropology in South Asia*. Ann Arbor: University of Michigan, 2000.

Lovell, N., and K. A. R. Kennedy. "Society and Disease in Prehistoric South Asia." In *Old Problems and New Perspectives in the Archaeology of South Asia*, edited by J. M. Kenoyer: 89–92. Wisconsin Archaeological Reports, vol. 2, Department of Anthropology, University of Wisconsin, 1989.

McIntosh, J. R. *A Peaceful Realm. The Rise and Fall of the Indus Civilization*. Boulder, Colo.: Westview, 2002.

Parpola, A. *Deciphering the Indus Script*. Cambridge: Cambridge University Press, 1994.

**You Are What You Eat (pp 15–20)**

Ambrose, S. H. "Effects of Diet, Climate and Physiology on Nitrogen Isotope Abundances in Terrestrial Foodwebs," *Journal of Archaeological Science* 18 (1991): 293–317.

Jerardino, A., J. Sealy, and S. Pfeiffer. "An Infant Burial from Steenbokfontein Cave, West Coast, South Africa: Its Archaeological, Nutritional, and Anatomical Context," *South African Archaeological Bulletin* 55 (2000): 44–48.

Sealy, J. "Stable Carbon and Nitrogen Isotope Ratios and Coastal Diets in the Later Stone Age of South Africa: A Comparison and Critical Analysis of Two Data Sets," *Ancient Biomolecules* 1 (1997): 131–147.

Sealy, J., et al. "Hunter-Gatherer Child Burials from the Pakhuis Mountains, Western Cape: Growth, Diet, and Burial Practices in the Late Holocene," *South African Archaeological Bulletin* 55 (2000): 32–43.

Van der Merwe, N. J. "Carbon Isotopes, Photosynthesis, and Archaeology," *American Scientist* 70 (1982): 596–606.

**The Lapedo Child (pp 21–23)**

Duarte, C., et al. "The Early Upper Paleolithic Human Skeleton from the Abrigo do Lagar Velho (Portugal) and Modern Human Emergence in Iberia," *Proceedings of the National Academy of Sciences* 96 (1999): 7604–7609.

Zilhão, J. "The Lagar Velho Child and the Fate of the Neanderthals," *Athena Review* 2 (4, 2001): 33–39.

Zilhão, J., and E. Trinkaus (eds.). *Portrait of the Artist as a Child: The Gravettian Human Skeleton from the Abrigo do Lagar Velho and its Archeological Context*. Vol. 22, *Trabalhos de Arqueologia*. Lisbon: Instituto Português de Arqueologia, 2002.

**The Moundville Dwarf Burials (pp 24–26)**

Powell, M. L. *Status and Health in Prehistory: A Case Study of the Moundville Chiefdom*. Washington, D.C.: Smithsonian Institution Press, 1988.

Snow, C. E. *Two Prehistoric Indian Dwarf Skeletons from Moundville*. Museum Paper no. 21, Alabama Museum of Natural History. Birmingham: University of Alabama, 1943.

**Buried with the Friars (pp 27–31)**

Arrizabalaga, J., J. Henderson, and R. French. *The Great Pox: The French Disease in Renaissance Europe*. New Haven, Conn., and London: Yale University Press, 1997.

Baker, B., and G. J. Armelagos. "On the Origin and Antiquity of Syphilis: A Palaeopathological Diagnosis and Interpretation, *Current Anthropology* 29 (5, 1988):703–737.

Daniell, C. *Death and Burial in Medieval England, 1066–1550*. London and New York: Routledge, 1997.

Evans, D. "Buried with the Friars," *British Archaeology* 53 (June 2000): 18–23.

Evans, D. H. *Excavations at the Augustinian Friary, Hull*. East Riding Archaeological Society monograph, forthcoming.

Evans, D. H., and K. Steedman (eds.). "Recent Archaeological Work in the East Riding," *East Riding Archaeologist* 9 (1997): 156–158.

Kolman, C., A. Centurion-Lara, S. A. Lukehart, D. W. Owsley, and N. Tuross. *Journal of Infectious Diseases* 180 (1999): 2060–2063.

Litten, J. *The English Way of Death: The Common Funeral Since 1450*. London: Robert Hale, 1991.

Power, C. "The Spread of Syphilis and a Possible Early Case in Waterford," *Archaeology (Ireland)* 6 (1992): 20–21.

Roberts, C. "Treponematosis in Gloucester, England: A Theoretical and Practical Approach to the Pre-Columbian Theory." In *L'origine de laf syphilis en Europe*, edited by O. Dutour, G. Palfi, J. Berato, and J.-P. Brun. "Avant ou après 1493?" Proceedings of a conference of the Centre Archéologique du Var, 25–28 November 1993. Paris: Errance, 1994.

Santos, F. R., A. Pandys, and C. Tyler-Smith. "Reliability of DNA-Based Sex Tests," *Nature Genetics* 18 (1998): 103.

Waldron, H. A., G. M. Taylor, and D. R. Rudling. "Sexing of Romano-British Baby Burials from the Beddingham Roman Villa, East Sussex," *Sussex Archaeological Collections* 137 (1999): 71–79.

Waldron, T. "DISH at Merton Priory: Evidence for a 'New' Occupational Disease?" *British Medical Journal* 291 (1985): 1762–1763.

**Lewis Man: A Face from the Past (pp 32–35)**

Cowie, T. G., and I. MacLeod. "Lewis Man: A Face from the Past," *Scots Magazine* 144 (1996): 426–428.

Cowie, T. G., and I. MacLeod. "Norman Helmet Interviews 'Lewis Man,'" *Young Archaeologist* 90 (1996): 5.

Dunwell, A. J., T. Neighbour, and T. G. Cowie. "A Cist Burial Adjacent to the Bronze Age Cairn at Cnip, Uig, Isle of Lewis," *Proceedings of the Society of Antiquaries of Scotland* 125 (1995): 279–288, fiche 2:G1–10.

MacLeod, I., and T. Cowie. "A Face from the Past," *Current Archaeology* 147 (1996): 100–101.

MacLeod, I., and T. G. Cowie. "A Face from the Bronze Age," *Society of Antiquaries of Scotland Newsletter* 7.2 (1996): 1.

**C18: The Pounder from Ur (pp 36–39)**

Hall, H. R., and C. L. Woolley. *Al-'Ubaid*. Vol. 1 of *The Ur Excavations*. Oxford University Press, 1927.

Hamilton, W. J. *Textbook of Human Anatomy*, 2nd ed. London: Macmillan, 1976.

Lockhart, R. D. *Living Anatomy*. London: Faber and Faber, 1974.

Warwick, R., and P. L. Williams. *Gray's Anatomy*, 35th ed. London: Longman, 1973.

**The Einiqua: People of a Hot, Dry Frontier Land (pp 40–44)**
Morris, A. G. "The Einiqua: An Analysis of the Kakamas Skeletons." In *Einiqualand: Studies of the Orange River Frontier*, edited by A. B. Smith. Cape Town, South Africa: University of Cape Town Press, 1995.
Morris, A. G. *The Skeletons of Contact. A Study of Protohistoric Burials from the Lower Orange River Valley, South Africa*. Johannesburg, South Africa: Witwatersrand University Press, 1992.
Morris, A. G., and A. L. Rodgers. "A Probable Case of Prehistoric Kidney Stone Disease from the Northern Cape Province, South Africa," *American Journal of Physical Anthropology* 79 (1989): 521–527.

**Bodies from the Ashes: Herculaneum and Pompeii (pp 46–50)**
"Anglo-American Project in Pompeii," *Archaeology* magazine website: www.archaeology.org.
Channel 4 (UK) website: www.channel4.com/plus/pompeii/volcano.htm.
Gore, R. G. "After 2000 Years of Silence, the Dead Do Tell Tales at Vesuvius," *National Geographic* 165 (1984): 556–613.
Nappo, S. *Pompeii*. London: Weidenfeld and Nicolson, 1998.
Sigurdsson, H., S. Cashdollar, and R. S. J. Sparks. "The Eruption of Vesuvius in AD 79," *National Geographic Research* 1.3 (1985): 332–387.
Steffy, J. R. "The Herculaneum Boat: Preliminary Notes on Hull Details," *American Journal of Archaeology* 89 (3, 1985): 519–521.

**The Wife of the Marquis of Dai (pp 51–55)**
Buck, D. "The Han Dynasty Tomb at Ma-wang-tui," *World Archaeology* 7 (1975): 30–45.
Fu Juyou and Chen Songcheng. *Mawangdui Han mu wenwu. (The Cultural Relics of the Han Dynasty tomb of Mawangdui)*. Changsha, 1992.
Hunan shen bowuguan (Hunan Provincial Museum), ed. *Mawangdui yi hao Han mu*. Beijing: Wenwu Publisher, 1973.
Hunan Yixueyuan (Medical Academy of the Hunan Province), ed. "Changsha yi hao Han mu gu hu yanjiu conghe baogao." In *Mawangdui Hanmu yanjiu. (A Study of the Corpse of the Han Dynasty tomb 1 of the Mawangdui)*. Changshe: Hunan renmin chubanshe, 1979.
Loewe, M. *Ways to Paradise: The Chinese Quest for Immortality*. London: George Allen and Unwin, 1979.
Pirazzoli-'t Serstevens, M. *China zur Zeit der Han-Dyanstie: Kultur und Geschichte. (Han Dynasty China: Culture and History)*. Stuttgart, 1982.
Silbergeld, J. "Mawangdui, Excavated Materials, and Transmitted Texts: A Cautionary Note," *Early China* 8 (1982/2): 83.
Wu Hong. "Art in a Ritual Context: Rethinking Mawangdui," *Early China* 17 (1992): 111–144.

**The Triple Burial at Dolní Vestonice (pp 56–58)**
Klíma, B. "Das jungpaläolithische Massengrab von Dolní Vestonice," ("The Upper Palaeolithic Common Cave of Dolní Vestonice"), *Quartär* 37/38 (1987): 53–62.
Klíma, B. "Une triple sépulture du Pavlovien à Dolní Vestonice," ("A Tribal Burial in the Pavlovian Dolní Vestonice"), *L'Anthropologie* 91 (1987): 329–334.
Klíma, B. *Dolní Vestonice II: Ein Mammutjägerrastplatz und seine Bestattungen. (Dolní Vestonice II: An Upper Palaeolithic Resting Place and its Burials.)* The Dolní Vestonice Studies, vol. 3, Études et Recherches Archéologiques de l'Université de Liège 73 (1995).

**Fenghuangshan Tomb 168, Jiangling, Hubei Province (pp 59–64)**
Chen Zhenyu. "Jiangling Fenghuangshan 168 hao Han mu" ("The Han Tomb of Fenghuangshan 168, Jiangling District"), *Kaogu* (1993–1994): 455–513.
Wu, Zhongbi, ed. *Jiangling Fenghuangshan 168 hao Han mu Xi Han gu shi yanjiu (Research on the Western Han Dynasty Mummy of Tomb 168 in Fenghuangshan, Jiangling District)*. Beijing: Wenwu Publisher, 1982.

**Turkana Boy: A 1.5-million-year-old Skeleton (pp 65–70)**
Walker, A., and R. Leakey (eds.). *The Nariokotome Homo erectus Skeleton*. Cambridge: Harvard University Press, 1993.

**Vilnius and the Ghosts of the Grand Armée (pp 71–76)**
Bertaud, J.-P., and W. Serman. *Histoire militaire de la France*. Paris: Éditions Fayard, 1998.
Rothenberg, G. E. *Les guerres napoléoniennes (The Napoleonic Wars)*. Paris: Autrement, 2000.

**Kennewick Man (pp 77–82)**
Chatters, J. C. *Ancient Encounters: Kennewick Man and the First Americans*. New York: Simon and Schuster, 2001.
Thomas, D. H. *Skull Wars: Kennewick Man, Archaeology, and the Battle for Native American Identity*. New York: Basic Books, 2000.

**The Iceman Reveals Stone Age Secrets (pp 84–90)**
Bortenschlager, S., and K. Oeggl (eds.). *The Iceman and his Natural Environment*. Vienna and New York: Springer Verlag, 2000.
Fleckinger, A., and H. Steiner. *The Fascination of the Neolithic Age: The Iceman*. Vienna-Bolzano: Folio Verlag and the South Tyrol Museum of Archaeology, 1999.
Fowler, B. *Iceman: Uncovering the Life and Times of a Prehistoric Man Found in an Alpine Glacier*. New York: Random House, 2000.
South Tyrol Museum of Archaeology website: http://www.archaeologiemuseum.it.
Spindler, K. *The Man in the Ice: The Discovery of a 5000-Year-Old Body Reveals the Secrets of the Stone Age*. New York: Harmony Books, 1994.

**The Butchered Anasazi (pp 91–94)**
Bullock, P. Y., et al. *Deciphering Anasazi Violence*. Santa Fe, N.M.: HRM Books, 1998.
Turner, C. G. II, and J. A. Turner. *Man Corn: Cannibalism and Violence in the Prehistoric Southwest*. Salt Lake City: University of Utah Press, 1999.
White, T. D. *Prehistoric Cannibalism at Mancos 5MTUMR-2346*. Princeton, N.J.: Princeton University Press, 1992.

**A War Monument in Gaul (pp 95–97)**
Brunaux, J.-L. (ed.) "Ribemont-sur-Ancre (Somme): bilan préliminaire et nouvelles hypothèses," *Gallia* 56 (1999): 177–284.
Brunaux, J.-L. *Les Religions Gauloises*. Paris: Errance, 2000.
Centre Archeologie Departemental de Ribemont-sur-Ancre website: www.ribemontsurancre.cg80.fr

**Windeby Girl: An Iron Age Bog Body (pp 98–102)**
Brothwell, D. *The Bog Man and the Archaeology of People*. London: British Museum Publications, 1986.
Coles, B., and J. Coles. *People of the Wetlands: Bogs, Bodies and Lake-Dwellers*. London: Thames and Hudson, 1989.
Coles, B., J. Coles, and M. Schou Jørgensen (eds.). *Bog Bodies, Sacred Sites and Wetland Archaeology*. Exeter: Department of Archaeology, University of Exeter, 1999.

Glob, P. V. *The Bog People. Iron-Age Man Preserved.* Ithaca, N.Y.: Cornell University, 1969.

van der Sanden, W. *Through Nature to Eternity: The Bog Bodies of Northwest Europe.* Amsterdam: Batavian Lion International, 1996.

**Who – or What – Killed Tutankhamen? (pp 103–107)**

Brier, B. *The Murder of Tutankhamen.* London: Weidenfeld and Nicolson, 1998.

Carter, H., and A. C. Mace. *The Tomb of Tutankhamen.* New York: Cooper Square, 1963.

Desroches-Noblecourt, C. *Tutankhamen: Life and Death of a Pharaoh.* New York: New York Graphic Society, 1963.

Frayling, C. *The Face of Tutankhamun.* London: Faber and Faber, 1992.

Reeves, N. *The Complete Tutankhamun: The King, the Tomb, the Royal Treasure.* London: Thames and Hudson, 1990.

Tyldesley, J. A. *Judgement of the Pharaoh: Crime and Punishment in Ancient Egypt.* London: Weidenfeld and Nicolson, 2000.

**The Cap Blanc Lady (pp 108–113)**

Archambeau, J., and P. Bahn. "Comment la Dame de Cap Blanc est arrivée à Chicago," *Bulletin de la Société Historique et Archéologique du Périgord* 128 (2001): 163–78.

Capitan, L., and D. Peyrony. "Trois nouveaux squelettes humains fossiles," *Revue anthropologique 22e année* 11 (November 1912): 4.

Dahlberg, A. A., and V. M. Carbonell. "The Dentition of the Magdalenian Female from Cap Blanc, France," *Man* 61 (48, March 1961): 2.

Field, H. "Cap Blanc Rock Shelter," *Antiquity* 12 (1938): 88–89, Plate III.

Field, H. "The Early History of Man with Special Reference to the Cap Blanc Skeleton," *Field Museum of Natural History Anthropology Leaflet* no. 26 (1927): 19.

Field, H. *The Track of Man: Adventures of an Anthropologist.* London: Peter Davies, 1955.

von Bonin, G. "The Magdalenian Skeleton from Cap Blanc in the Field Museum of Natural History," *Medical and Dental Monographs, University of Illinois (Urbana)* I (1935): 1–76.

**Batavia's Graveyard (pp 114–118)**

Dash, M. *Batavia's Graveyard: The true story of the mad heretic who led history's bloodiest mutiny.* London: Weidenfeld & Nicolson, 2002.

Drake-Brockman, H. *Voyage to Disaster.* Nedlands, Western Australia: UWA Press, 1995 (1963).

Edwards, H. *Islands of Angry Ghosts.* London: Hodder and Stoughton, 1966.

Godard, P. *The first and last voyage of the Batavia.* Perth: Abrolhos Publishing, 1993.

Pasveer, J. Buck, A. & van Huystee, M. "Victims of the Batavia Mutiny: physical anthropological and forensic studies of the Beacon Island skeletons." *Bulletin of the Institute for Maritime Archaeology* 22: 45–50, 1998.

Stanbury, M (ed). "Abrolhos Islands Archaeological Sites: interim report." *Australian National Centre of Excellence for Maritime Archaeology* Special Publication No.5, 2000.

**High-mountain Inca Sacrifices (pp 119–123)**

Douglas, K. "High Society," *New Scientist* 172 (2320, 2001): 30.

Reinhard, J. "At 22,000 Feet, Children of Inca Sacrifice Found Frozen in Time," *National Geographic* 196 (5, November 1999): 36–55.

Reinhard, J. "Research Update: New Inca Mummies," *National Geographic* 194 (1, July 1998): 128–135.

**Prehistoric Homicide and Assault (pp 124–129)**

Morris, A. G., A. I. Thackeray, and J. F. Thackeray. "Late Holocene Human Skeletal Remains from Snuifklip, near Vleesbaai, Southern Cape," *South African Archaeological Bulletin* 42 (1987): 153–160.

Morris, A. G., and J. E. Parkington. "Prehistoric Homicide: A Case of Violent Death on the Cape South Coast, South Africa," *South African Journal of Science* 78 (1982): 167–169.

Pfeiffer, S., N. J. Van der Merwe, J. E. Parkington, and R. Yates. "Violent Human Death in the Past: A Case from the Western Cape," *South African Journal of Science* 95 (1999): 137–140.

**The Sacrifices at Huaca de la Luna (pp 130–134)**

Benson, E. P. and A. G. Cook (eds.). *Huaca de la Luna Sacrifices: Ritual Sacrifice in Ancient Peru.* Austin: University of Texas Press, 2001.

**Pit of the Bones (pp 136–139)**

Arsuaga, J. L., et al. "Three New Human Skulls from the Sima de los Huesos Middle Pleistocene Site in Sierra de Atapuerca, Spain," *Nature* 362 (1993): 534–537.

Bahn, P. G. "Treasure of the Sierra Atapuerca," *Archaeology* (1996): 45–48.

Bermúdez de Castro, J. M., et al. *Atapuerca, Nuestros Antecesores.* Salamanca, Spain: Junta de Castillón y León, 1999.

Carbonell, E., et al. "La revolución de Atapuerca," *Revista de Arqueología* 19 (210, October 1998): 14–24.

Cervera, J., et al. *Atapuerca, un Millón de Años de Historia.* Madrid: Plot Ediciones, 1998.

**A Woman from Roman London: in a Lead Coffin (pp 144–147)**

Archaeotrace website: www.archaeotrace.co.uk

Frere, S. S., M. Roxan, and R. S. O. Tomlin (eds.). *The Roman Inscriptions of Britain II: Instrumentum Domesticum (Personal Belongings and the Like).* Fascicule 1. Gloucester, England: Alan Sutton Publishing, 1990.

Kleiner, D. E. E., and S. B. Matheson (eds.). *I, Claudia: Women in Ancient Rome.* New Haven, Conn.: Yale University Art Gallery, 1996.

Prag, J., and R. Neave. *Making Faces: Using Forensic and Archaeological Evidence.* London: British Museum Press, 1997.

Richards, J. *Meet the Ancestors: Unearthing the Evidence That Brings Us Face to Face with the Past.* London: British Broadcasting Corporation, 1999.

"Roman Britain in 1999," *Britannia* 31 (2000): 420 (preliminary report).

Swain, H., and M. Roberts (eds.). *The Spitalfields Roman,* 2nd ed. London: Museum of London, 2001.

**The Romito Dwarf (pp 148–150)**

Frayer, D. W., R. Macchiarelli, and M. Mussi. "A Case of Chondrodystrophic Dwarfism in the Italian Late Upper Paleolithic," *American Journal of Physical Anthropology* 75 (1988): 549–565.

Frayer, D. W., W. A. Horton, R. Macchiarelli, and M. Mussi. "Dwarfism in an Adolescent from the Italian Late Upper Palaeolithic," *Nature* 330 (1987): 60–62.

**Anne Mowbray and the Skeletons in the Tower (pp 151–154)**

Brook, A. H. "A Unifying Aetiological Explanation for Anomalies of Human Tooth Number and Size," *Archives of Oral Biology* 29 (1984): 373–378.

Demirjian, A., H. Goldstein, and J. M. Tanner. "A New System of Dental Age Assessment," *Human Biology* 45 (1973): 211–227.

Louda, J., and M. MacLagan. *Lines of Succession.* London: Orbis, 1981.

Molleson, T. "Anne Mowbray and the Princes in the Tower: A Study in Identity," *London Archaeologist* 5 (1987): 258–262.

Puech, P.-F., and F. Cianfarani. "La Paléodontologie: Étude des maladies des dents," *Dossiers de l'Archéologie* 97 (1985): 28–33.

Rushton, M. A. "The Teeth of Anne Mowbray," *British Dental Journal* 119 (1965): 355–359.

Schour, I., and M. Massler. *Journal of the American Dental Association* 28 (1941): 1153.

Tanner, L. E., and W. Wright. "Recent Investigations Regarding the Fate of the Princes in the Tower," *Archaeologia* 84 (1934): 1–26.

Warwick, R. "Anne Mowbray: Skeletal Remains of a Medieval Child," *London Archaeologist* 5 (1986): 176–179.

**The Mysterious Burials of the Okunev Culture (pp 155–159)**

Gryaznov, M. *The Ancient Civilization of South Siberia*. London: Barrie and Rockliff, 1969.

Savinov, D., and M. Podolski (eds.). *Okunevski sbornik (The Okunevo Book): Culture, Art, Anthropology* (in Russian). St. Petersburg: Petro-RIF, 1997.

Vadetskaya, E. B., N. V. Leontjev, and G. A. Maksimenkov. *The Monuments of the Okunev Culture* (in Russian). Leningrad: Nauka, 1980.

**Positioned for Political Influence (pp 160–164)**

Fash, W. *Scribes, Warriors, and Kings: The City of Copán and the Ancient Maya*. London: Thames and Hudson, 1991.

Martin, S., and N. Grube. *Chronicles of Maya Kings and Queens*. London: Thames and Hudson, 2000.

McAnany, P. A. *Living with the Ancestors: Kinship and Kingship in Ancient Maya Society*. Austin: University of Texas Press, 1995.

**Chinchorro Mummies (pp 166–170)**

Arriaza, B. T. *Beyond Death: The Chinchorro Mummies of Ancient Chile*. Washington, D.C.: Smithsonian Institution Press, 1995.

Pringle, H. *The Mummy Congress: Science, Obsession, and the Everlasting Dead*. New York: Hyperion, 2001.

**The Mysterious Mummy in Tomb 55 pp 171–174)**

Aldred, C. *Akhenaten, King of Egypt*. London: Thames and Hudson, 1988.

Reeves, N., and R. H. Wilkinson. *The Complete Valley of the Kings*. London: Thames and Hudson, 1996.

Tyldesley, J. A. *Nefertiti: Egypt's Sun Queen*. London: Viking, 1998.

Tyldesley, J. A. *Private Lives of the Pharaohs*. London: Channel 4 Books, 2000.

Tyldesley, J. A. *The Mummy*. London: Carlton Books, 1999.

**Restoring the Royal Mummies (pp 175–178)**

Harris, J. E., and E. F. Wente (eds.). *An X-ray Atlas of the Royal Mummies*. Chicago: University of Chicago Press, 1980.

Harris, J. E., and K. R. Weeks. *X-Raying the Pharaohs*. London: Macdonald, 1973.

Smith, G. E. *The Royal Mummies*. Cairo: Service des Antiquités de l'Egypte, 1912.

**Funerary Rituals of the Tashtyk Culture (pp 179–185)**

Kyzlasov, L. R. *The Tashtyk Epoch in the History of the Khakas-Minusinsk Basin* (in Russian). Moscow: Moscow University Press, 1960.

Vadetskaya, E. B. *Les idoles anciennes de l'Ienissei* (in Russian and French). Leningrad: Nauka, 1967.

Vadetskaya, E. B. *The Tashtyk Epoch in the Ancient History of Siberia* (in Russian). St. Petersburg: Peterburgskoe Vostokovedenie, 1999.

# Picture Credits

The publisher would like to thank the following for permission to reproduce their images. While every effort has been made to ensure this listing is correct, the publisher apologises for any omissions or errors.

**Bohuslav Klima:** p1, 45(cr), 56, 58. **David Frayer:** p2, 135(cr), 148(t), 149(b), 150(b). **South Tyrol Archaeological Museum:** p3(l), 83(cl), 84, 85, 86, 87, 88, 89, 90. **Institute of Archaeology, Portugal:** p3(r), 9(l), 21, 22, 23. **Richard Reisz:** p3(c), 103, 104(b), 105, 106, 165(r), 171, 173, 174, 176, 177, 178. **Eurelios Agence de Presse:** C.Munoz: p5(cl), 6(b), 83(r), 98, 99, 100, 101, 102; UMR 6578/CNRS/: p45(cl), 73, 74, 75, 76, 77; C Pouedras: p95, 96, 97; P Plailly: p165(l), 166, 167, 168, 169, 170. **Trevor Cowie/Iain Macleod:** p5(cr), 9(cl), 32, 33, 34, 35. **Art Archive:** p5(l), 10, 45(cr), 47, 48, 49. **Museum of London:** p5(r), 135(r), 144, 146, 147(t). **Johan Reinhard:** p5(c), 6(t), 119, 120, 121, 123. **Dept. of Archaeology and Museums, Govt. of Pakistan © J.M. Kenyer:** p7, 11, 13, 14. **Agence France Presse:** p8. **Natural History Museum/Theya Molleson:** p9(cr), 36, 37(t), 38(b), 39(b). **Professor Alan G. Morris:** p9(r), 40(b), 41, 42, 43, 44, 126(b), 127, 128, 129. **Illustrated London News:** p12. **Dr. Antonieta Jerardino, University of Cape Town:** p15, 17. **Judith Sealy:** p18, 19, 20. **University of Alabama Museums:** p24, 25. **Ohio Historical Society:** p26. **Humber Archaeology Partnership:** p27, 28, 29, 30, 31. **Cape Town Archives Repository:** p40(t) AG7146/95. **Margarete Pruech:** p45(cl), 60, 61, 64. **NASA:** p46. **National Geographic Magazine:** p50. **Anthropos Institute, Moravian Museum:** p57. **Alan Walker © National Museum of Kenya:** p65, 66, 67, 68, 69. **Collection of the Gilcrease Museum, Tulsa:** p82(b). Black Hawk and His Son Whirling Thunder by John Wesley Jarvis; registration no. 0126.1007. **Tri-city Herald:** p78, 81(t). **James Chatters:** p80, 81(b), 82(t). **Griffith Institute, Oxford:** p83(l), 104(t), 107. **Field Museum Chicago:** p83(cr), 112, 113. **Peter Bullock:** p91, 93. **Powerstock:** p92, 94, 160, 161, 162, 163, 164. **Paul Bahn:** p109, 110, 111, 148(b), 149(t), 150(t). **Western Australian Maritime Museum:** p114, 115, 116, 117. **Susan Pfeiffer:** p124, 125, 126(t). **Steve Bourget, Department of Art and Art History, Texas University:** p130, 131, 132, 133, 134. **Madrid Scientific Films:** p135(l) 136, 137, 138, 139. **Vladimir Bazaliiskii:** p135 (cl), 140, 141, 142, 143. **Archaeotrace Limited:** p145, 147(b). **Elena Miklashevich:** p155, 156, 157, 158, 159, 165(cl), 179, 180, 181, 182, 183, 184, 185.

The line drawings on pp 18 and 20 have been reproduced with the permission of the South African Archaeological Society.

The image on p 39(t) previously appeared in A Richards: *Land, Labour and Diet in Northern Rhodesia*. London: Oxford University Press, 1939.

The images on pp 51–55 previously appeared in Fu Juyou and Chen. Songcheng. *Mawangdui Han mu wenwu. (The Cultural Relics of the Han Dynasty tomb of Mawangdui)*. Changsha, 1992, and Hunan sheng bowuguan (Hunan Provincial Museum), ed. *Mawangdui yi hao Han mu*. Beijing: Wenwu Publisher, 1973. The items are currently exhibited at the Hunan Provincial Museum, Changsha, China.

The images on pp 59–64 show items currently exhibited at the Jingzhou Museum, Jingzhou City, China.

Anne Thackeray would also like to acknowledge Alan G. Morris of the Department of Human Biology at the University of Cape Town, and J. Francis Thackeray of the Transvaal Museum.

# Contributors

**Paul Bahn**: *Consultant Editor*
TOPICS: The Lapedo Child: p 21–23; The Triple Burial at Dolní Vestonice: p 56–58; A War Monument in Gaul: p 95–97; The Cap Blanc Lady: p 108–113; Pit of the Bones: p 136–139; The Romito Dwarf: p 148–150.

**Bernadette Arnaud**
Journalist for French science magazine *Sciences et Avenir*, and *Archaeology* magazine (New York).
TOPIC: Vilnius and the Ghosts of the Grand Armée: p 71–76.

**Vladimir Bazaliiskii**
Department of Archaeology, Irkutsk University.
TOPIC: The Prehistoric Graves of Siberia: p 140–143.

**Caroline Bird**
Expert in Aboriginal and Heritage Studies, Independent Archaeologist.
TOPIC: *Batavia*'s Graveyard: p 114–118.

**Peter Bogucki**
Associate Dean, School of Engineering and Applied Science, Princeton University.
TOPICS: The Iceman Reveals Stone Age Secrets: p 84–90; Windeby Girl: An Iron Age Bog Body: p 98–102.

**Peter Bullock**
Director, Archaeological Investigations of Chicago.
TOPICS: The Moundville Dwarf Burials: p 24–26; Kennewick Man: p 77–82; The Butchered Anasazi: p 91–94.

**Trevor Cowie**
Department of Archaeology, National Museum of Scotland.
**Mary Macleod**
Western Isles of Scotland regional archaeology specialist.
TOPIC: Lewis Man: A Face from the Past: p 32–35.

**Dave Evans**
Manager of the Humber Archaeology Partnership.
TOPIC: Buried with the Friars: p 27–31.

**David Gill**
Sub-Dean, Faculty of Arts and Social Studies, University of Wales Swansea.
TOPICS: Bodies from the Ashes: Herculaneum and Pompeii: p 46–50; A Woman from Roman London: In a Lead Coffin: p 144–147.

**Patricia McAnany**
Professor of Archaeology, University of Boston.
TOPIC: Positioned for Political Influence: p 160–164.

**Jane McIntosh**
Freelance archaeological writer and lecturer.
TOPIC: The Mohenjo Daro "Massacre": p 10–14.

**Elena Miklashevich**
Department of Archaeology, Kemerovo University.
TOPICS: The Mysterious Burials of the Okunev Culture: p 155–159; Funerary Rituals of the Tashtyk Culture: p 179–185.

**Theya Molleson**
Department of Palaeontology, Natural History Museum.
TOPICS: C18: The Pounder from Ur: p 36–39; Anne Mowbray and the Skeletons in the Tower: p 151–154.

**Margarete Prüch**
Research Associate KAVA, Bonn (DAI)
TOPICS: The Wife of the Marquis of Dai: p 51–55; Fenghuangshan Tomb 168, Jiangling, Hubei Province: p 59–64.

**Anne Thackeray**
Research Associate in Archaeology, University of the Witwatersrand
TOPICS: You Are What You Eat: p 15–20; The Einiqua: People of a Hot, Dry Frontier Land: p 40–44; Turkana Boy: A 1.5-million-year-old Skeleton: p 65–70; Prehistoric Homicide and Assault: p 124–129.

**Joyce Tyldesley**
Honorary Research Fellow in Archaeology, University of Liverpool.
TOPICS: Who – or What – Killed Tutankhamen?: p 103–107; The Mysterious Mummy in Tomb 55: p 171–174; Restoring the Royal Mummies: p 175–178.

**Karen Wise**
Associate Curator of Anthropology, Natural History Museum of Los Angeles.
TOPICS: The Sacrifices at Huaca de la Luna: p 130–134; High-mountain Inca Sacrifices: p 119–123; Chinchorro Mummies: p 166–170.